"Although innovation at John Deere has been core for over a century, the insights and practical concepts of *Payback* are helping us link innovation more rapidly and directly to value creation."

—Robert W. Lane, Chairman and CEO, Deere & Company

"Nonprofits seeking innovative solutions to the world's most difficult problems can benefit from the concepts and tools provided in *Payback*. Philanthropies can benefit greatly by paying attention to the four S factors of payback—start-up, speed, scale, and support costs—when making their decisions about awarding grants and supporting new projects."

—Linda Segre, member of the Executive Team, Google.org

"How can you turn research into new products and services that truly generate a cash return, and how can you do it faster? *Payback* offers a fundamental reassessment of these critical questions and shows how sales and science can reinforce each other. This is an important book for anyone who wants to make their innovative activities more profitable."

—Amar Bhidé, Lawrence D. Glaubinger Professor of Business, Columbia University, and author of *The Origin and Evolution of New Businesses*

"The authors detail what is important to get fast payback from innovation. They provide insight into the many factors required, especially how to lead an innovative company and its people and how to align an organization to create an innovative climate."

—Claus Weyrich, Member of the Managing Board and Head of Corporate Technology, Siemens AG

Payback

Payback

Reaping the Rewards
of Innovation

James P. Andrew
Harold L. Sirkin

with John Butman

Harvard Business School Press
Boston, Massachusetts

Library of Congress Cataloging-in-Publication Data

Andrew, James P., 1962-
  Payback : reaping the rewards of innovation / James P. Andrew, Harold L. Sirkin,
with John Butman.
    p. cm.
  Includes bibliographical references and index.
  ISBN-13: 978-1-4221-0313-5 (hardcover : alk. paper)
  1. Technological innovations—Management. 2. Industrial management. I. Sirkin,
Harold L. II. Butman, John. III. Boston Consulting Group. IV. Title.
  HD45.A628 2006
  658.5'14—dc22

                                                              2006024530

The paper used in this publication meets the minimum requirements of the American
National Standard for Information Sciences—Permanence of Paper for Printed Library
Materials, ANSI Z39.48-1992.

To our families

Carole, Nicholas, and Christine

—Jim

Eileen, Jessica, and David

—Hal

CONTENTS

Payback seeks to help solve the most important problem that confronts businesspeople today: how to get a better return on their investment in innovation.

We've worked with hundreds of companies, in virtually every industry and in countries around the world. Most of them are expert at what they do, and many of them are world leaders. Nevertheless, we have seen only a few that are operating at, or even close to, their potential when it comes to achieving a return on their investment in innovation. As a result, they are unable to maximize their shareholder value, grow at a rate they would like, or respond as effectively as they could to competitive pressure.

Many companies are looking for organic growth (after more than a decade of seeking growth primarily through mergers and acquisitions, the Internet, and other avenues) and believe that innovation is the engine that can drive it for them. But a large percentage of the companies that set their sights on innovation fail to make a full commitment to it. One reason for this is that innovation, unlike cost cutting or merger and acquisition activity or regional expansion, is hard to define and quantify. As a result, many companies waste a great deal of their spending on innovation. If they could improve the return by even 10 or 20 percent, it could make a huge difference in virtually every aspect of their corporate performance.

Some executives have also told us that they can't innovate successfully because their companies don't generate enough good ideas.

But, in our experience, lack of ideas is rarely the issue. Thousands of good ideas exist within every organization, even those that don't think of themselves as innovative. The real problem these companies have is how to turn their ideas into cash. They have not developed a process for collecting the ideas, screening them, nurturing them, and then commercializing and realizing them in a way that achieves payback.

We believe that any company can innovate and achieve a healthy return if it deliberately and consistently does the following: sets clear goals for its innovation efforts, operates in a disciplined way, selects the optimal innovation business model for each new product or service, aligns its organization around innovation, and exercises leadership practices that encourage, motivate, and enable the people within the company to innovate.

If a company does generate cash through innovation, it can create new ideas and products and services and processes, and achieve organic growth for itself. In turn, it can help stimulate growth in the global economy and increase the quality of everyday life for its employees, customers, and people around the world. Innovation can create new markets and help economies adapt to changing environments—by developing whole new approaches to such pressing concerns as energy, health care, education, and poverty.

We wrote *Payback* to help readers who have experience with innovation but are not yet satisfied with their return, as well as for those who are turning their attention to innovation for the first time. We have tried to strike a balance between the theoretical and the practical—the book offers both a structure for thinking about innovation and concrete ideas and examples for managing and executing it. Both are illustrated with stories we have gathered from a wide variety of companies around the world, and supported by the insights and ideas of their leaders.

We have learned a great deal from our field work over the past twenty-five years, and we've had the privilege to engage with some of the best practitioners of innovation and the finest minds in many business endeavors. In 2003, we decided to quantify our experience and conducted our first annual, large-scale survey of the views and practices of executives involved in innovation. In 2006, we enlarged the scope of the survey still further and partnered with *BusinessWeek* in its analysis and publication. The results of this unique BCG/*BusinessWeek* "Senior Management Survey on Innovation" have greatly added to our knowledge and informed the writing of the book.

As a result of this combination of long experience and extensive qualitative and quantitative research—shaped and clarified by the activities of analysis and writing—we have come to believe more deeply than ever that innovation is far more than a discrete product, service, or improvement. It is a holistic process and a personal and corporate journey—and payback is central to it.

And because profitable innovation is so important to individual companies and to the global society, you, as a leader, cannot begin the journey too soon.

ONE

Overview

F OR ALMOST EVERY COMPANY, the greatest challenge of innovation is not a lack of ideas but rather, successfully managing innovation so that it delivers the required return on the company's investment of money, time, and people. Most attempts at innovation fail to deliver this return—they do not generate enough payback.

Payback means one thing—cash. Cash that is realized within the planned time frame. When a company makes an investment in innovation and creates something new that produces a cash return swiftly and directly, it has created a winning situation, particularly when the return is larger than expected. The company has a "hit" on its hands. And this is true regardless of whether the new thing is a product, service, process change, business model, customer experience, or anything else that is new.

But it is the nature of innovation, of all types, that cash is not always produced from it, and rarely is it produced immediately. There can be a lag between the time of investment in innovation and the

cash return. This lag can make companies and leaders nervous. Perhaps the cash payback will never come at all? With other types of
investments (particularly in tangible assets like factories, machines,
or new trucks), companies can often calculate their cash return with
much more certainty. But, as with advertising and certain other expenditures, the return on an investment in innovation cannot be so easily
predicted or measured.

To complicate matters, the innovation process sometimes generates a cash payback, but indirectly—not through the specific product
or service being developed but through a benefit that only later impacts
the company's ability to generate cash. These indirect benefits are real,
although difficult to capture. There are four of them:

- **Knowledge.** The innovation process always produces knowledge, some of which can usually be put to work in more than
  one way to produce cash.

- **Brand.** Innovation can enhance a brand, thereby attracting
  more customers and enabling companies to charge a premium,
  which can mean greater cash returns.

- **Ecosystem.** Innovators can create exceptionally strong
  ecosystems of partners and associated organizations, enabling
  them to leverage their position in multiple ways, for the benefit of their payback.

- **Organization.** People want to work for and contribute to innovative companies, and being innovative allows companies to
  attract and retain more of the best people, or at least more of
  the most innovative ones. Having better people, with less cost
  to keep them, results in more cash.

For managers, the fundamental challenge of innovation is to achieve
the required cash payback, by managing the overall innovation process

with the understanding that payback can come quite directly and quickly, but also that it may take longer, be much less certain, or come back to the company only indirectly, via other products and services.

To achieve payback, companies must manage the innovation process holistically and with discipline. They must make careful choices about how much and where to invest. They need to be smart about which innovation business model they choose to execute with. And they must deliberately align and lead their organizations toward payback.

They must also accept that innovation—more than other business strategies—entails a significant amount of risk. There are three types of risk: technical, operational, and market. If the new product or service has some technical failings, if the organization cannot actually commercialize or realize it, or if the market does not embrace the product as planned, the company is put at risk of not achieving the needed or desired payback.

Many companies try to remove as much of the risk as possible—installing strict procedures and ironclad approval mechanisms. There is value in control, of course, but for the most part, trying to make innovation risk-free either stifles the process or causes people to lower their sights, so nothing big ever happens. A few companies, surprisingly, are "risk optimistic." They embrace risk too freely and end up paying for it. Companies must learn to understand risk, how to analyze and evaluate it, and how to manage it. And they must realize that often the greatest risk for a company seeking to grow is to take no risk at all.

When companies manage innovation with this understanding and in these ways, they can create new products and services (as well as process improvements, customer insights, new business models, and other types of innovations) that deliver payback and increase the company's ability to continue to grow and thrive. Failure to do so puts the company on the path to commoditization, nondifferentiated positions, lack of advantage—and lack of cash.

Many companies are not achieving the payback they would like. This is clear both from our experience in the field and from the results of our annual survey—The Boston Consulting Group (BCG)/*Business Week* "Senior Management Survey on Innovation," which was first conducted by The Boston Consulting Group (BCG) in 2003.

In this survey, completed in April 2006, 1,070 executives representing 63 countries and all major industries responded, and they answered 19 questions.[1] The bottom-line finding was that 48 percent of the respondents said they are not satisfied with the return they get on their investments in innovation, and they gave a fascinating array of reasons for their dissatisfaction. Here are just a few of the most typical:

- "We make exaggerated estimates of the benefits of a new product."

- "We have not established satisfactory performance metrics that consider leading, as well as lagging, financial factors."

- "We pursue too many things simultaneously and cannot execute them all."

- "We don't have the right people or capacity in place."

- "Our time to market is too slow."

- "Our sales force focuses on our traditional business."

- "Seniors managers won't fund new products because they are too risky."

- "Innovation is not a priority for the board of directors."

- "We have a block in mind-set."

All of these issues—and almost all of the others that we gathered in the survey—pertain to issues of management, capabilities, metrics, mind-set, decision making, and leadership. Very few respondents to the survey said that their problem with innovation was a lack of ideas. This is significant, because there has been a great deal of attention paid in recent years to issues that have to do with ideas—including practical methods of idea generation, the importance of creativity, and the role of invention.

But as important as ideas are, they are only a small part of the issue. Creativity is not synonymous with innovation. Good, even great, ideas are not enough to guarantee payback. Innovation is the entire process of developing ideas *with the goal of achieving payback*, and it comprises three phases of activity, each with a distinct output.

- **Idea generation.** This is the phase when ideas are generated, developed, tested, evaluated, and refined, but during which the company makes no commitment to actually creating a product or service (or taking some other action) based on the ideas. The output of this phase is . . . an idea.

- **Commercialization.** This phase begins with the green light from management to develop a proposed idea into an offering that can be produced and marketed, either externally or internally, and ends when the product is launched to the buying audience. At this point, the process has produced an *invention*—the technology, product, service, or process improvement that has achieved a tangible form but that has not yet been tested by the external (or internal) market. The invention is the thing itself. Innovation is the process.

- **Realization.** This phase begins with market launch and ends when the product or service comes to the end of its life cycle. Although this is the phase when the cash payback is achieved, many important aspects of the size and timing of the payback have been determined by the activities of the earlier phases.

As our survey suggests—and as we've found in the vast majority of companies we've worked with—a lack of ideas is not the main stumbling block to achieving payback from innovation.

The central challenge for innovators, and the main focus of this book, is the commercialization phase. This is true whether the idea is for a new product, service, business model, customer experience, or any other new thing. It is during this phase that the company must evaluate the payback potential of the available ideas. It must determine the proper level of investment for each idea. It must choose which innovation business model to use to develop and produce the product. It must figure out how to organize the company to innovate and achieve payback. And its leaders must find the best way to lead the effort.

Payback is organized into three parts, the main ideas of which are summarized in the remainder of this chapter:

1. **What Is Payback?** Chapters 2 and 3 explain the four S factors that directly affect cash payback, as well as the indirect benefits of innovation that can lead back to cash.

2. **Choosing the Optimal Model.** Chapters 4 through 6 explore the characteristics, advantages, and management challenges of the three innovation business models—integration, orchestration, and licensing—and how they affect risk and payback.

3. **Aligning and Leading for Payback.** Chapters 7 and 8 discuss how to align all key elements of the company around innovation, and the essential elements of leadership that are needed to achieve maximum payback from innovation.

The Characteristics of Cash

In innovation, as we've said, cash truly is king. Four factors have a direct impact on cash payback:

- *Start-up costs*, or prelaunch investment

- *Speed*, or time to market

- *Scale*, or time to volume

- *Support costs*, or postlaunch investment, which includes a variety of costs and continuing investments

These *S* factors can be visually expressed in the cash curve, shown in figure 1-1. The curve graphically plots cumulative cash flow over time. It makes clear many of the managerial challenges and assumptions that often get hidden when looking at spreadsheets of annual cash flows and projections. Net present values, various option valuations, and multiple scenario analyses are useful (and valuable), but discussing and debating the shape of the cash curve will make these financial projections much more sound.

As we'll discuss in chapter 2, the cash curve is a tool that is extremely helpful in decision making, planning, analysis, and communication. Most important, consistent use of the cash curve can help executives more effectively manage innovations for cash payback.

FIGURE 1-1

The cash curve

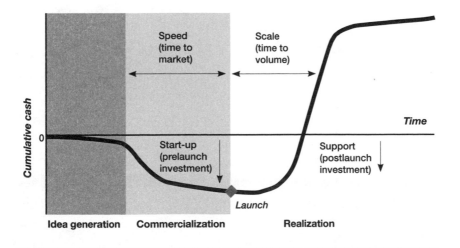

The Four S Factors That Directly Affect Cash

Start-up costs, or prelaunch investment. The size of the prelaunch investment is also known colloquially as the "size of the hole"—that is, how deep the curve falls below the zero line. Many companies don't fully track the amount of their investments in innovation by project and, as a result, are not able to determine the payback of a specific new product or service. A large prelaunch investment may enable a company to develop assets and capabilities that will result in a large cash payback, but it also increases the marketplace success required to generate payback, and will affect how the innovation process is managed and which innovation model is chosen.

An example of a new product with a very large prelaunch hole is Iridium, the consortium created by Motorola to develop the first globally operative mobile phone. The prelaunch investment was about $5 billion, which might not have been excessive, had the other parts of the cash curve played out differently. Unfortunately, the Iridium system did not work properly, subscribers stayed away in droves, and the venture was eventually terminated and assets sold for $25 million.

Speed. Increasing speed and reducing time to market can increase payback by enabling a company to capture a larger market share at a higher average selling price and by starting the cash flow quickly. Yet an overly aggressive time to market may disproportionately increase development costs, reduce the quality of the innovation, or have a negative impact on one of the indirect benefits. (For example, rushing to market may increase payback but incur a cost of excessive project team turnover.)

Being faster to market also may come with a higher cost in educating the customer about the new product—and if the efforts are not successful, the first-mover advantage may be lost. TiVo, for example, pioneered the digital video recorder (DVR) but has struggled for years to get enough customers to understand exactly what its product can do. Now that the DVR is growing more popular, the electronics giants are introducing their own models, and it looks increasingly unlikely that TiVo will ever achieve payback.

Scale. Time to volume is the time it takes from launch until the new product achieves volume production on the scale planned and that can deliver payback. A company can control its ability to supply the product, and it can seed market demand, but it cannot dictate market acceptance. Ideally, the time-to-volume part of the cash curve is short and steep. The faster a new product reaches scale, the quicker it can begin to generate a substantial amount of cash payback. However, a steep time to volume may also put a strain on the organization and its supply chain, increasing costs and potentially reducing cash returns.

Microsoft's development and commercialization of its Xbox video game consoles offers a good illustration of the issues that are involved in achieving scale. When Microsoft launched its first-generation Xbox in 2001, the company was trailing far behind Sony. The Xbox had to reach a critical mass of sales soon after it launched in order to persuade game developers to write new games for it. That need drove Microsoft's decision about which innovation business model to choose, as we'll see in chapter 5.

Support costs, or postlaunch investment. To maximize the payback from each innovation once it is in the market, a company must determine how much to invest in a product or service, for how long, and in what areas. These decisions about support often are affected by the innovation business model and the way the organization is aligned.

When Whirlpool introduced its Gladiator GarageWorks garage organization system, for example, the company decided to introduce the wall panels at a very low price to the consumer—so low that the company would make little profit on these units. The wall panels, however, were designed to hold a wide variety of other components of the Gladiator system—including racks and shelving, baskets, bins, and hooks. Although consumers did not need the wall panels to own Gladiator components, they were much more likely to buy more components when they had the panels in place. Selling the wall units at a very low margin was essentially a postlaunch investment designed to increase cash payback for the entire Gladiator product line.

There are many other types of support costs, including marketing and promotional activities, product improvements and refreshments, and sales, distribution, and channel initiatives.

Cash Traps

As we have seen over and over again in businesses around the world, many companies do not carefully analyze or manage the four S factors that directly affect payback for each new product. As a result, they often overestimate the payback potential or are unable to realize it. The reality (and often hidden truth) is that many new products and services, even seemingly successful ones, don't achieve any cash payback over their entire lifetimes. They actually cost more to create and support than they deliver in cash return. These are *cash traps*.

Sometimes a company will deliberately continue to make and market a product that is actually a cash trap, because it delivers one of the indirect benefits of innovation in a significant way. But this is rare. Most companies keep cash traps in their portfolio without realizing they are doing so. Cash traps drain resources and can even become large enough to destroy an entire company.

Polavision, for example, Polaroid's excursion into instant film-based movies, was a cash trap that helped bring about the company's demise. Polavision was a technical tour de force when launched in 1977, but it required such a large prelaunch investment and was so slow getting to market that videotape beat it to the punch. Polavision was dead soon after launch, and Polaroid eventually had to write off its $197 million in inventory. We'll discuss cash traps in more detail in chapter 2.

Indirect Benefits That Can Lead to Cash

Payback from innovation means cash, but innovation also delivers four other noncash benefits that can be valuable when they have the potential to lead to cash—indirectly or over time—but with some certainty.

Knowledge

In the process of creating new products and services, a company always generates knowledge, in the form of patents, copyrights, trade secrets, trade dress, and other intellectual assets.

As we'll discuss in chapter 3, acquiring knowledge can take time and cost money, but it can also create value and contribute to payback even beyond its specific and immediate application. Knowledge can often be applied to other new products and services, or sold or licensed as a product in and of itself. The management challenge is to determine how great an investment to make in knowledge acquisition, according to how much it will contribute to current and future payback.

Companies often pursue projects primarily for the knowledge benefit rather than to directly generate cash. Sony created its famous electronic dog, AIBO, for example, primarily as a way to learn more about robotics. Teruaki Aoki, senior executive vice president and executive officer at Sony, told us, "I really encouraged the people in the AIBO group to push hard, because all of the basic technology that was used in the AIBO—the sensors, and the AI, and so on—could eventually be applied to Sony's major products."[2] AIBO operated using software and electronics that Sony believes will become increasingly important in creating customizable, human-responsive consumer electronics products for many years to come—a few of which are already beginning to appear. The knowledge benefit clearly can be converted, over time, into cash payback.

Brand

The fact that a company is innovative can enhance and strengthen its brand, simply because it is seen as being innovative. A company that is perceived as innovative can build stronger relationships with partners and suppliers, and make customers more accepting of new products and services and more willing to pay higher prices for them—all of which contribute to payback.

In 2002, LG Electronics Digital Appliance Company introduced to the U.S. market a 26-cubic-foot titanium-clad "Internet refrigerator" that retailed for $8,000. With computer capabilities, an Ethernet connection, and a 15-inch LCD television set into the door, the refrigerator enabled homeowners to check e-mail, surf the Web, and watch TV. At the time, LG was considered a supplier of low-end goods, primarily known for its Goldstar brand name. The new refrigerator was well received, but the size of the market for a fridge-PC was tiny. For LG, however, that wasn't the point. As CEO Young-Ha Lee, president of LGE's Digital Appliance Company, explained to us, the Internet refrigerator greatly benefited the LG brand. "The market was surprised when we attacked the premium, high end of the market," Lee told us. "Now people see LG as a younger, more innovative company."[3] As a result, LG has not only increased its overall share in the U.S. market, it has also been able to compete successfully in the premium segment. The brand benefited from innovation, which boosted cash payback.

Ecosystem

Innovative companies can often create special relationships with members of their ecosystem—including customers, suppliers and collaborators, channels, investors and shareholders, the press, analysts, and regulators.

These relationships can bring a number of advantages that lead to cash. For example, innovative companies can gain preference with suppliers and partners, which will give them first consideration over their competitors or better terms in their deals and contracts. Innovative companies can also achieve exclusivity at retail, giving them wider exposure, better placement, and more volume than companies that are not seen as innovative. And innovative companies can create ecosystem relationships that enable them to establish a standard. Dolby Laboratories, for example, worked closely with the professional sound-recording market to develop a breakthrough noise reduction system. These professional connections helped Dolby establish its technology as the standard in noise reduction systems for the consumer market, which has led to enormous cash payback for the company.

Organization

Innovative companies are more able to attract, retain, and enable people who have particular skills in the creation of new products and services—as individual participants, team members, or leaders—than other companies are, which often means they can manage the process of innovation more effectively and generate higher payback.

Companies sometimes create new products and services primarily to benefit the people within their own organizations. In 2002, for example, The Boston Beer Company, home of Samuel Adams Boston Lager, launched a beer called Utopias—a beer with the highest concentration of alcohol in the world, an almost unbelievable 25 percent, in comparison to the 5 or 6 percent of a typical beer. The product brought new energy to the company and, according to founder Jim Koch, helped institutionalize innovation at Boston Beer—thus leading to the creation of other new products that have generated cash payback. "This product may be marginally profitable at best," Koch told us, "but the benefit comes in another way—pride in creating something that has never been done before. For our company, which is focused on patience and pride, plus a relentless dissatisfaction with the status quo, this is really important."[4]

The Innovation Business Models

An innovation business model determines how an idea is commercialized and realized. Each model is a distinct approach to the choice of activities the company and its ecosystem participants undertake, as well as to how cash, indirect benefits, and risk are allocated. The three approaches are not the same as strategies—such as first-mover and fast-follower—nor are they ownership structures, like joint ventures and strategic alliances, although they can be used alongside them. And the models extend far beyond discrete processes such as new product development or product life cycle management, but they incorporate them.

There are three innovation business models:

- Integration

- Orchestration

- Licensing

Today, most innovative large companies use all three innovation business models in various combinations, especially when they have diverse portfolios of products, and change them as company and market conditions warrant.

Unfortunately, many companies employ a model by default rather than by carefully selecting the one that is best for the given invention. They simply go along with the model they have always used, that others are using, or that is easiest for them at the moment. As a result, they don't get the cash payback or the indirect benefits that they might otherwise, because the default model does not suit the innovation or because they simply aren't able to manage the execution.

It's important to make a deliberate choice of model, because the choice almost always has a significant impact on three elements of payback:

- The probability of success in achieving a return

- The amount of payback

- The allocation of payback and risk between different parties during commercialization and realization

Each model has advantages and disadvantages. The model needs to be chosen to best suit the nature of the idea itself, the capabilities of the company, the resources available, and how it will affect the cash curve factors and indirect benefits. Each innovation business model can deliver payback, but each presents risks and trade-offs that must be managed.

The Integrator

Integration is the most familiar innovation model and, historically, the dominant one. As we'll discuss in chapter 4, the integrator "owns" and manages the entire innovation process. The most compelling reason to integrate is that the company wants to exert strict control over the S factors that affect cash or key indirect benefits, or it wants to keep these positive results for itself. With the prospect of higher cash return or greater indirect benefit, however, comes higher risk.

Companies also integrate for specific operational reasons that can affect payback, such as when quality must be tightly controlled, when a company needs to move very quickly to market and partners can't be counted on to supply key capabilities or to meet important deadlines, or when the company wants to keep close ownership of the knowledge associated with the product.

Although many companies choose to integrate because they seek the greatest control over the innovation process, integration requires world-class skills—such as in manufacturing, marketing, and sales— and true cross-functional cooperation to succeed. It is very possible for an integrator to lose control of some part of its own process or to fail completely.

In chapter 4, we'll see how Seagate, the world's largest maker of disk drives, has distinguished itself from competitors with its commitment to integration in an environment where most competitors are pursuing a different innovation business model as their primary vehicle to generate payback.

The Orchestrator

Orchestrators control and manage every aspect of innovation but do not execute all of them, as we'll see in chapter 5. Companies orchestrate when they lack the capabilities required to create a new product and, for various reasons, don't wish to acquire the capabilities themselves. Orchestration can provide more flexibility than integration, because the company does not have to commit itself to personnel, capital equipment, organization structures, or markets that might need to be changed during the life cycle of the product or service. Orchestration is also a model companies use when they need to shake up their thinking or inject outside ideas into the process. Open innovation is a form of orchestration that companies primarily use in the idea generation phase.

The orchestrator approach usually requires a lower investment than does integration. Companies can draw on the assets or capabilities of partners, and the orchestrators' own assets and capabilities contribute to only part of the process. For example, many automotive original equipment manufacturers (OEMs) are contracting out design and production activity to so-called coach builders in Europe. Companies like Valmet, in Finland; Magna Steyr, in Austria; Pininfarina, in Italy; and Karmann in Germany reduce OEM costs and eliminate management layers by enabling the parceling up of various design and production activities. These companies help form the ecosystem that enables orchestration in the automotive industry.

Managing orchestration is very different from managing integration, because the relationships are more collaborative than they are hierarchical. In chapter 5, we'll see how Microsoft turned to orchestration to create the Tablet PC.

The Licensor

Although orchestration and integration are the most widely used innovation business models, licensing is rapidly becoming a favored route to payback, as we'll discuss in chapter 6. A licensor is the primary owner of the spark of the new product and sometimes of its commercialization, but has no ownership of the realization. Even so, some licensors specify exactly how their intellectual assets are to be used, to ensure certain standards of quality, performance, and consistency of brand (if their brand name is involved). In such cases, it's almost as if the licensor is "renting" the business system of its licensee, so as to avoid the cost and effort involved in bringing its idea to market. Some licensors develop close relationships with their licensees, so they can take advantage of new knowledge gained through realization and apply it to further improvements.

Licensing is widely used in industries such as biotech and information technology, where the pace of technological change is rapid and risks are high. Smart companies see licensing as a way to improve the cash curve, better leverage scarce resources, and take advantage of the capabilities of other companies.

Dolby Laboratories, for example, deliberately chose to license its knowledge assets and build a strong brand with certain audiences, and it gained tremendous payback. When Dolby went public in 2005, the IPO generated almost a half-billion dollars for its founder on the first day of trading, proof of just how lucrative licensing can be.

Aligning

A major reason that companies do not achieve payback is that they think of and manage innovation as an ad hoc activity or something that can only be encouraged but not institutionalized. Such efforts can sometimes result in an occasional success, but will never enable a company to create a stream of new products and services that consistently achieve payback. To do that, alignment is necessary.

Alignment means that a company's business strategy, processes, organization structure, business model, people, metrics and rewards, and leadership are all geared toward innovation. No innovative company has more complete alignment than BMW, and we'll describe the three-level alignment process developed and rigorously followed by the German automaker in chapter 7.

Alignment can be achieved within all manner of organizational structures. There is no ideal model of layers, units, and spans of control. But, as we'll discuss in chapter 7, all the companies we have worked with and explored pay close attention to certain elements of their organization that must be aligned. These include the following.

Individual Responsibility

Although no one arrangement of boxes and lines will lead to an improved capability to innovate, most innovative companies do build innovation responsibility into the organizational reporting structure. They have either a facilitator, who encourages and advocates for innovation, or a chief innovator, who is essentially the innovation leader and gets directly involved in the process. We'll discuss how companies create individual responsibilities for innovation throughout their operations.

Unit Responsibility

Companies often establish small groups or units with very specific roles in the innovation process. Some of these units are focused on the creation of a single product or service, and tend to operate outside the normal operations of the company. This is primarily so they can escape the "tyranny of the P&L." Many projects with good payback potential are scratched or abandoned by managers with direct responsibility for a P&L because their cash return will come too far in the future (which for some means beyond the end of the current fiscal year).

Giving the responsibility for these small innovation units to managers who are not driven by a quarterly P&L often gives new projects a much better chance of surviving and eventually generating payback (and helping some future general manager hit his numbers). Some of these units are incubators, designed to encourage, seek out, evaluate, and promote ideas. Others operate like internal venture capitalists or sponsors, selecting and funding ideas and pushing them through the commercialization phase. Motorola's Early Stage Accelerator (ESA) is just one example of such a small, innovation-focused internal unit.

Companywide Responsibility

Dedicated innovation leaders and innovation-focused groups play important roles, but it's also essential that everyone within the company understand and believe in the importance of the innovation process and be able to make a contribution to it. This is more than a matter of providing ways for nonmanagers and line employees to make suggestions for improvements or contribute ideas; it is about developing a language of innovation that everyone can speak—as Samsung has done—and aligning all their activities around innovation, as BMW and other innovators have succeeded in doing.

Conducive Conditions

Certain organizational and cultural conditions can be created that are conducive to idea generation and that motivate people to be innovative. As we'll see in chapter 7, there are six such conditions that leaders can impact: time to think, space to explore, deep domain knowledge, stimulation, an idea-focused dialogue, and motivation.

Openness

More and more, companies recognize that they are operating within a network, and usually a network of networks, and that they must take advantage of knowledge and expertise that do not exist within the boundaries of their organization. They build relationships, create structures, and cultivate networks that increase the diversity of voices, ideas, influences, and practices in their innovation process. In chapter 7, we'll see how some companies, like Schindler, the innovative producer of elevators, become scouts—always on the search for new ideas from the outside. And how others, like Procter & Gamble (P&G), become beacons—attracting people to bring them new ideas and technologies.

Measurement

Companies often rely on a number of traditional measures—including R&D spending as a percentage of sales, number of patents filed, and percentage of sales from new products and services (often defined as those that have been introduced within the previous two to three years)—to determine or prove how innovative they are. But none of these measures has much to say about payback. Companies that achieve payback carefully measure the four factors that directly affect the cash curve and cash payback (start-up costs, speed, scale, and support costs) and constantly monitor the indirect benefits that can lead to future payback. We'll explore how some companies, such as Philips, have created custom metrics to better track their return on innovation.

Leading

Innovation requires a type of leadership that is very different from that needed to pursue a strategy of geographic expansion, cost cutting, or acquisition. Many leaders who have little experience with creating new products and services assume that the management of the innovation process can be delegated and, in fact, needs to be delegated, because it's not their area of expertise.

But innovation cannot be handed off or passed down; leaders must be involved. This does not necessarily mean that the leaders must personally manage or even participate in any specific activity. Rather, it means that they spend a meaningful percentage of their time genuinely

engaging with all phases of the process, put sufficient emphasis on innovation in their communications and actions, and devote appropriate
resources to each phase.

As we'll see in chapter 8, there is a short list of actions and decisions that have significant impact on payback and that the leadership
must not allow to be taken by default:

- **Convincing an organization that innovation matters.** The leader
  needs to convince people within the company that innovation
  is fundamentally important, to the company, to them, and also
  to the leader himself.

- **Allocating resources.** When many projects—innovative or
  otherwise—are vying for scarce resources of time, money,
  and talent, it is up to the leader to decide how they should
  be distributed.

- **Choosing an innovation business model.** The leader must
  ensure that the model is deliberately selected to optimize
  payback for each product and service rather than being "chosen" by default.

- **Reshaping dynasties.** Often, when a product or service has
  been extremely successful, it will continue to get high allocations of resources, even after its payback begins to decline. It
  becomes a "dynasty" and can throttle innovation in other parts
  of the company. To reduce the negative influence of a dynasty,
  leaders must take purposeful action, often with regard to how
  resources are allocated to the dynasty or to other business
  units that might affect the dynasty.

- **Focusing on the right things.** The leader is the one who focuses
  the organization on the ideas and activities that will contribute
  to cash payback. This often means choosing which ideas to
  commercialize, how to configure the portfolio, and when to kill
  an idea or cannibalize an existing product.

- **Assigning the right people to the right role.** Leaders must take responsibility for hiring or assigning the people who can most help the innovation process, and also for removing or reassigning those people who cannot help or are getting in the way.

- **Dealing with risk.** Leaders of successful innovators recognize that to achieve payback and, in addition, to create a legacy of note, sparks must fly and risks must be taken. Rather than just support or pay lip service to the innovation effort, they must truly lead it.

In the afterword, we address a topic of great interest to readers who wish to apply the ideas in this book to their businesses: the concrete steps they can take to get started.

But first, we suggest that you begin the journey to payback by gaining a deeper understanding of the pivotal role that cash must play in successful innovation.

PART ONE

What Is Payback?

I N CHAPTER 2, Cash and Cash Traps, we discuss the paramount importance of cash payback and the four *S* factors—start-up costs, or prelaunch investment; speed, or time to market; scale, or time to volume; and support costs, or postlaunch investment—that affect it. We introduce the cash curve, a tool that helps executives visualize, discuss, analyze, improve, and set goals for payback.

In chapter 3, The Indirect Benefits of Innovation, we discuss the four indirect benefits of innovation—knowledge acquisition, brand enhancement, ecosystem strength, and organizational vitality—and how they can, and must, link back to cash payback.

TWO

Cash and Cash Traps

Profitability is the only decisive indicator for
being innovative.

—Claus Weyrich, member of the managing board of
  Siemens AG, head of corporate technology

JUDGING BY CASH PAYBACK, Microsoft is the most successful innovator in business history. Microsoft Windows
and Microsoft Office not only have transformed the world, they have
delivered billions of dollars in cash return. Even today, two decades
after its introduction to the market, Windows generates about $1 bil-
lion a month in revenue and some $9 billion annually in operating in-
come, or more than $750 million every month.[1] Windows has probably
generated the largest payback of any new product ever, including such
blockbusters as the Model T Ford, the Boeing 747, and the cholesterol-
lowering drug Lipitor, one of the most successful new medicines ever
launched.

Not only have Windows and Office delivered a huge payback to Microsoft, the cash has helped the company achieve remarkable growth. The numbers, while well-known, are still staggering. Founded in 1975, Microsoft has increased its revenue in every year of its existence and earnings in all but two of those years. From 1985 to 2004, revenue increased at a compound annual growth rate of 34 percent while earnings increased at 36 percent annually.

But Microsoft has not allowed Windows to become an innovation-thwarting "dynasty." The company has continued to innovate around its core product, integrating it with other services and technologies, adding new features and capabilities, and bringing it to new platforms.

Microsoft has also demonstrated that it knows Windows won't be its payback engine forever. It has devoted enormous resources to developing and commercializing new technologies and products—such as MSN, Xbox, and the Tablet PC—that could become the company's next major source of payback.

Many readers will no doubt strongly question (as did some of the people we showed early versions of this manuscript to) the thought of Microsoft as the most innovative company of all time. After all, Microsoft didn't invent computing or the user interface. But that's irrelevant. As Microsoft has proved, successful innovation does not always rest on original invention. Microsoft built on an idea that it did not generate, then innovated around it, and has achieved cash payback more successfully than any other company in history.

What matters above all else in the case of Microsoft and its success in innovation is cash payback—not only because public companies in much of the world have a fiduciary responsibility to generate cash payback, but because cash is what enables a company to create organic growth and keep on innovating. Shareholders may tolerate short periods of lower profitability or slower growth, but only if they believe that the company is working on something that will eventually deliver a large cash payback. Unless an innovation generates cash, it is, in reality, an expense.

So, the reason to innovate is simple—to generate cash.

The Cash Curve

In order to manage for cash, it's necessary to have a disciplined and consistent way to analyze, understand, and make decisions about how to manage it—and the most effective tool for doing so is the cash curve.

Although most companies fill plenty of spreadsheets with masses of financial data about their innovation projects, the information contained in them is hard to apply to the innovation process. It's sterile, hard to connect to decisions, and so voluminous as to be virtually impossible to use effectively. Accordingly, "running the numbers" becomes an exercise for finance jockeys, and holistic business discussion and decisions become secondary to "making the numbers work." While it is important to have detailed information about cash flow, net present value, and financial outcomes of multiple scenarios, these tools all too frequently don't enable different groups of people within a company to visualize the implications and alternatives that would help them make effective business decisions.

The cash curve, however, forces managers to think through the dynamics of cash, helps them clearly see what will happen as the result of investment and management decisions, enables them to identify sensitive areas, and always provokes discussion about how to "improve the curve." This is a case where a picture is worth more than a thousand words—it may be worth $100 million or more.

The cash curve brings a discipline to the discussion that is hard to achieve when talking about new products and services that, obviously, have no prior performance data to go on. It enables the project managers and executives to see how trade-offs in the size and timing of investments will affect cash payback. The data that is used to create the curve need not be perfect (how could it be?), but it can constantly be refined as the project continues.

Without this kind of discussion, managers often enter the innovation process with very different assumptions about the goals of the effort, wildly divergent attitudes toward risk, and no shared approach to managing the process. But with the cash curve and the discussion it generates, the leaders can get everyone on board to set goals, identify risks, and align the various parts of the organization around the effort.

The discussion focuses on the four S factors that affect the success of a new product or service and its ability to generate payback. These factors are usually hidden in spreadsheets, but the cash curve (see figure 1-1 on p. 7) makes them visible and the trade-offs much more explicit.

- Start-up costs, or prelaunch investment

- Speed, or time to market

- Scale, or time to volume

- Support costs, including reinvestment

Start-up Costs

The first S factor is the size and timing of the prelaunch investment, or "depth" of the curve beneath the cash "breakeven" line. A large start-up investment may enable a company to develop assets and capabilities that will result in a substantial cash payback, but it also increases the risk of the innovation. A large investment multiplies the marketplace success required to generate cash payback, and will affect how the innovation process is managed and which business model is chosen.

A classic example of a large upfront investment that did not pay off is the Iridium global mobile telephony network. When Motorola decided to develop the world's first mobile phone system that could work globally through a network of satellites, the company knew that it was embarking on a path of innovation that would require new elements in all three phases of the process: idea generation, commercialization, and realization. In order to share the considerable prelaunch investment of this new venture, Motorola put together a consortium of partners and created a separate entity, called Iridium. By doing so, the company hoped to keep its part of the prelaunch investment curve as shallow as possible and also leverage the capabilities of others.

Motorola was wise to take on partners, because the start-up curve cut much deeper than they had estimated—some $5 billion was spent during the prelaunch period. This might have been acceptable if Iridium had moved with greater speed. But in the twelve years it took from idea to launch, competitors were able to create their own mobile telephony systems that offered the services customers wanted at prices they were willing to pay.

And while the $5 billion and twelve years of effort produced a great deal—including sixty-six satellites launched into low-earth orbit, connection gateways around the world, and a $180 million marketing campaign—the expense was far more than the consortium had anticipated.[2] If they had drawn and carefully looked at a cash curve, they would have clearly seen the big hole they were digging for themselves, how important it would be to reach scale very fast, and how little cash return there would be to apply to support activities if scale was not quickly achieved.

FIGURE 2-1

Iridium cash curve

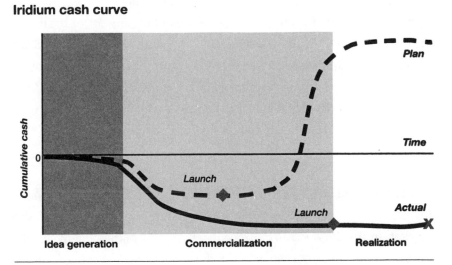

If Iridium's managers had been forced to acknowledge just how deep (and wide) the start-up hole was, and how sharply the scale part of the curve had to rise (see figure 2-1), they might also have better understood that there was very little margin for error with the product/service itself—it would have to work perfectly as promised, and a large percentage of the target market would have to accept it very quickly.

Unfortunately, neither happened. The Iridium phone had technical difficulties from the start. The handset was bulky and difficult to operate, and users had to find a clear patch of sky overhead to get a signal. In addition, the pricing was off. Airtime was too expensive, even for the businesspeople who were the primary target audience. No doubt the price was set high to gain as much cash return as possible, but it actually had the opposite effect—it limited the audience and increased the time it would take to achieve scale. Iridium had forecast that the network would be serving six hundred thousand customers by 2000, but by the end of 1999, it had signed up only fifteen thousand subscribers. Meanwhile, traditional wireless carriers were entering the market much more quickly than Motorola had expected. The combination was fatal.

The plug was pulled in August 1999, and Iridium went into bankruptcy. When the news came out, stories in the media advised people to watch the skies for abandoned satellites plunging out of orbit and disintegrating in the atmosphere.[3] (Innovation quite literally going down in flames!) Iridium's assets were eventually purchased by a group of investors for just $25 million.[4] But the development process for Iridium had an upside. It yielded significant knowledge for Motorola, which the company applied to other products that later achieved success.

Speed, or Time to Market

The second S factor that affects payback is speed—the time it takes from the "eureka" moment of idea generation to launch into the market. In the 2006 BCG/*BusinessWeek* "Senior Management Survey on Innovation," many respondents identified speed as a major problem at their companies. In fact, "the time it takes to develop an idea" was cited as the number-one barrier to increasing innovation payback.

Increasing speed to market can increase cash payback and decrease risk by enabling a company to capture a larger market share at a higher average selling price (especially as increased global competition accelerates the pace of commoditization) and decreasing the start-up costs.

Yet, when a company focuses on increasing speed that decision often has an impact on the other factors. Sometimes an overly aggressive time to market may increase (rather than decrease) start-up costs, or affect the quality of the product, which may reduce its ability to achieve scale.

The major goal of increasing speed is usually to reach the market before effective competitors can appear, and thus achieve maximum share of the market. But being faster to market may entail a higher support cost, because customers will need to be made aware of, and educated about, the new product. If the required investment cannot be made or the efforts are not successful, the advantage of speed may be lost.

Consider TiVo, the U.S. company that helped pioneer the digital video recorder (DVR), which connects cable and satellite TV systems to a device, giving viewers the freedom to record and manage shows. The DVR has become a huge hit, shaking up the broadcasting and advertising industries in the process. TiVo, however, has struggled for years to get people to understand exactly what a DVR can do, and now that it is growing more popular, the consumer electronics giants have begun to offer their own versions with enhanced features. Cable and satellite operators also offer their own versions at lower prices. TiVo may never achieve payback on its amazing invention (see figure 2-2).

Speed has been a major issue for more than two decades, and over the past ten years—as the life cycle for products of all kinds has shortened—speed has only become more important.[5] The shorter the expected life of the product category and the faster the likely competitive response, the more important it is not to be late to market. Categories are not as stable as they used to be. It's important that the product has

FIGURE 2-2

TiVo cash curve

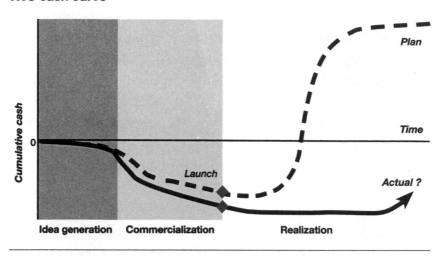

as much time as possible to achieve payback before the market moves on. As the postlaunch section of the curve gets shorter, the start-up cost must be spread out over a lower unit volume of sales. Speed, it seems, is getting faster.

Scale, or Time to Volume

Scale is the time it takes from launch until the new product or service achieves planned volume. A company can, to some extent, control its manufacturing capabilities, but it cannot dictate market demand. Ideally, the time-to-volume part of the curve is short and steep. The faster an innovation reaches full production volume, the quicker it can begin generating cash profits.

Microsoft's development and commercialization of its second-generation Xbox video game console offers a good illustration of the issues involved in managing the time it takes to achieve scale (see figure 2-3).

Microsoft launched its first-generation Xbox console in November 2001, at about the same time as Nintendo introduced its GameCube

FIGURE 2-3

Xbox 360 cash curve

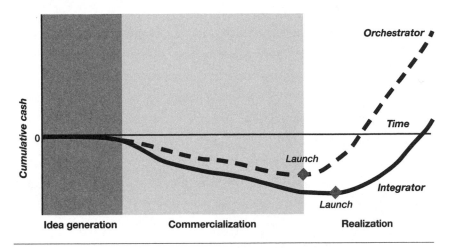

console. But Sony had launched PlayStation 2 about a year earlier, and by the time Xbox launched, PlayStation 2 had already established a user base of 6.5 million units. Microsoft had difficulty getting gamers to switch to its machine, and it also took a lot of work to persuade developers to create Xbox games.

With the second-generation Xbox, Microsoft was determined not to lag the market again. It knew that Sony planned to launch its PlayStation 3 in the first half of 2006, so Microsoft launched Xbox 360 in November 2005, hoping that its improved gameplay would entice game developers to create cutting-edge Xbox games and hard-core gamers to buy it. But that would only happen if Microsoft could reach a critical mass of sales quickly enough.

So, rather than manufacture the Xbox 360 itself, Microsoft chose to work with Flextronics, a contract designer and manufacturer of electronics hardware, just as it had on the original Xbox. Flextronics was joined by Wistron and Celestica, two additional contract manufacturers, and they jointly developed a plan for quick ramp-up, in order to be ready to build enough products to meet the hoped-for global demand. "The faster I can build volume, the better off I am," Robbie Bach, president of Microsoft's Entertainment and Devices Division, told the *Wall Street Journal*.[6]

Microsoft planned well, and also got lucky, because Sony announced that the launch of PlayStation 3 would be delayed until November 2006. This gives Microsoft an opportunity to sell approximately 10 million consoles before its rival's launch—which historically has been the amount needed for a machine to achieve critical mass and success in the video games industry.

It's entirely possible, however, that the installed base of PlayStation 2 gamers and developers will not immediately leap to the new Xbox, since Sony's marketing juggernaut is certain to position PlayStation 3 as the ultimate gaming console. To counteract Sony's marketing messages, Microsoft will need to make significant investments in education, promotion, marketing, and relationship building in order to achieve scale.

It's also possible that once PlayStation 3 is introduced, Microsoft could counterattack by slashing the price of Xbox 360 and introducing Halo 3, the latest installment of its best-selling first-person "shooter." These activities, if pursued, will further increase the company's total investment.

Support Costs

Investment in postlaunch support activities can have a significant effect on profit. These costs include:

- Marketing and promotional activities

- Pricing actions

- Product improvements and extensions

- Cannibalization of other products in the portfolio

When the cash curve intersects the breakeven line—the place where cumulative cash investment returns to zero and heads north—the new offering is on its way to achieving payback. However, given the time value of money, it will need to climb to some point above the breakeven line before it actually generates a cash return.

To maximize its return on the investment in the new product, the company now must determine how much to invest in supporting the product—through marketing activities, creating product enhancements and extensions, increasing distribution, and adjusting pricing. The flip side, of course, is determining when it makes sense not to invest further in supporting a product. That can mean deciding to cannibalize it with the launch of the company's own next-generation product or simply to let the product shrink in share, slowly decline, and eventually be phased out.

On the cash curve, the postlaunch profit of a new product will often look lower than it does when the product's performance is calculated by annual profit. Companies often support products because they appear to be delivering cash payback, although in actuality they are not. This is because the annual profit figure usually does not fully capture all of the true postlaunch costs—including not only cash but also support costs and the cost of scarce resources, particularly key staff. Another cost that is harder to assess, but can be significant, is the opportunity cost of not creating and launching other products because the resources have been committed to the product now on the market. While self-cannibalization is always difficult, it is even more painful to keep a product on the market too long and watch as its market share is gobbled up by competitors.

Using the Cash Curve

When developing a new product or service, managers usually think they know what will be involved in commercializing it, and they make various estimates about how it will perform once it enters the market. However, their assumptions are often based on past history (or wishful thinking) rather than on asking and objectively answering a series of questions about the process and the range of possible outcomes, such as:

- Can the team complete the development of the product on time in order to get it into production at the targeted date?

- Will the product really meet the customer-derived spec?

- Will our production process be able to get the hoped-for yield?

- Can we produce as many units as we think we can?

- How much capital investment will be required—and how might that fluctuate?

- When will we sell the first unit?

- How many units will customers buy? At what price?

- When are we going to go cash positive?

- In short, does this invention look as if it can generate cash payback? If so, when and how?

At Linde, the German maker of material-handling equipment, the cash curve of a new product is negative for the first couple of years (see figure 2-4), because that's how long it takes to develop a new truck, and the investment required is substantial. Even so, the management team looks for ways to improve the curve, and weighs the trade-offs that will affect the postlaunch profit over the long term.

As Stefan Rinck, a board member at Linde, told us, "For instance, the ramp-up to the scale we need can take quite a long time or it can take less, depending on what we do. We try to accelerate the ramp-up as much as possible, but only to a certain level, because when we're

FIGURE 2-4

Linde cash curve

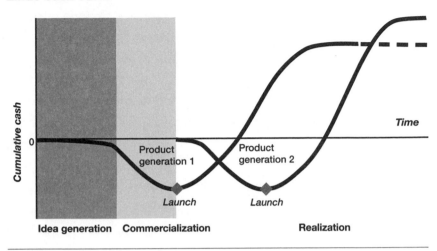

bringing out a new product, in most of the cases we also are phasing out an existing one. We have to watch the time between phasing out and ramping up, because otherwise we have two products to produce. That means twice as many pieces and parts for the factory to handle, which is very, very expensive. The longer the product is successfully in the marketplace, without any replacement, the better it is. So we try to keep a truck in production as long as possible, because that improves our cash flow. After ten or fifteen years, we usually see the return start to peak, and we know it's time to bring in a new product."[7]

Using the cash curve forces management to raise and discuss key questions and enables it to better analyze the risks and optimize the plan before making a commitment to invest and move forward. The cash curve forces managers to bring together all their different perspectives on an idea, with the goal of creating a cash curve that everyone can understand, support, and work to achieve. It offers a common point of reference for people in different positions and disciplines throughout the company and enables them to assess the performance of a new product or service throughout the innovation process.

Don Remboski, vice president of innovation at steel bearing maker Timken, is a strong believer in cash curve analysis. "It makes you stop and ask some difficult questions. You realize that if we miss our start date by a week, net present value will change a huge amount—maybe as much as millions of dollars per week. You realize that you need to have a great deal of confidence in the numbers you're projecting. That's when the business people start digging deeper. They go back to marketing and ask, 'Is the market really going to take X number of these things in week thirty-seven of 2007?' And the marketing people usually say, 'Hold on. We were just kind of guessing there. Let me take a little time to refine that number.' When you get the refined number back from marketing, you realize that the picture isn't quite as rosy as it seemed. For example, the number will only be good if you ship on a certain date and no later. Otherwise, the whole projection falls apart. Of course, the market is far more complex than you can ever model with complete accuracy, but you've got to do the best you can to make smart decisions about which products to pursue and which to leave alone."[8]

Most companies don't attempt to analyze the cash curve of an individual product or their entire portfolio. Typically, they put together a project profit and loss (P&L) statement that includes a net present value number, given all the assumptions that have been loaded into the analysis. This has the effect of focusing everyone's attention on the things they need to do to make the net present value number look good. That's the wrong thing to do.

The right thing to do is to use the cash curve to show what the payback will be if the assumptions you have made are right. This allows you to make a plan based on those assumptions, understand the impact of each assumption, determine which ones are most critical, test the ones with the biggest impact, and make trade-offs during commercialization. With this perspective, the process can be managed, not simply reacted to.

The cash curve is also a useful tool for analysis of projects that go astray, providing a framework for a discussion of what went wrong, when, and why—and how to avoid the same problems next time around.

The cash curve brings a steadying influence to the entire innovation process and all those involved in it. This is important because it's very easy to get excited about the prospects of a new idea, overestimate its potential for payback, and seriously underestimate the risks involved.

Assessing Risk

When companies do not use the cash curve to help them understand and evaluate the payback potential of a new idea, they often take on the wrong amount or wrong type of risk. They sometimes take such a large financial risk that they threaten the business's overall ability to continue to invest, and at the extreme, put the company in jeopardy. Or they are so unwilling to take on risk, usually because they overestimate their risk exposure, that they cannot produce growth from their ideas. Or they take on a type of risk that the company is unprepared for, usually in terms of capabilities. Understanding where risk can change the shape of the cash curve, and its impact on payback, allows companies to think more holistically about which projects are really risky, and which ones are less so.

"When we think about our innovations, and our broader portfolio, the main criterion is the financial reward—not the amount of sales, but the amount of margin, the amount of money we can get," said Pierre-Emmanuel Levy, deputy vice president of R&D at Saint-Gobain, a global manufacturer and provider of high-technology materials and associated services, founded in 1665. "The second aspect is risk. We try to have a unified method for all the hundreds of projects we have. And we also try to have a unified way to avoid the risk, which could be a commercial risk, a technical risk, and also a legal risk. For each project we try to be able to calculate this with a risk attractivity metric. This, combined with the financial reward, gives us the tools to really manage the portfolio."[9]

There are three main types of risk:

- **Executional.** Can the company actually develop, produce, distribute, and support the new product or service as scheduled?

- **Technical.** Will the product or service work or perform as intended or promised?

- **Market.** Will the product or service be accepted by customers in the amounts, at the prices, and within the time, desired?

The amount and kind of risk involved in the proposed idea have a fundamental impact on which innovation business model the company should use to realize and commercialize it. Unknown or untested methods, technologies, or practices may be required. It may not even be clear that the idea actually can be commercialized at all or that it will work as intended once it has been.

The risk of developing new products in the pharmaceutical industry, for example, is substantial and well known. When a new drug fails to achieve its potential, there can be large write-offs, often more than $10 million and occasionally as much as $100 million or more. In late 2004, Biogen Idec withdrew its multiple sclerosis drug Tysabri from the market, due to unexpected side effects in patients and a linkage to a rare brain disease.[10] Biogen Idec fell drastically in market capitalization from almost $20 billion in November 2004 to $15 billion in 2005.[11]

But risk can also be much more mundane and mechanical, and can occur long before a product comes to market. Cadbury Schweppes, for example, launched a new kind of gum, called Trident Splash, in 2005. It took two years and millions of dollars to develop the gum. It was one of the company's biggest new product developments ever, a three-layered concoction of candy, gum, and a liquid center. According to the *Wall Street Journal*, however, there were operational problems in early manufacturing runs—the machines crushed the pellets and the inner liquid oozed out, creating what are known at Cadbury as "leakers."[12] These problems lengthened the time to market for the new gum, which meant the company could not achieve scale as soon as it would have liked.

The use of the cash curve to evaluate the potential of an opportunity for payback—its size, type, and timing—will provide a perspective on risk. The three types of risk affect the curve in different ways, at different times.

Technical risk is largely a start-up issue (see figure 2-5). It can increase prelaunch investment and lengthen the time to market. Market risk determines the time to scale and the amount and timing of support costs. Executional risk can impact the entire curve. By looking at the curve, companies can change the way they manage risk in two ways.

First, it can make clear the trade-offs that can be made and their impact on the curve. Things can always be done faster, cheaper, or with less risk, but rarely can all three be accomplished at once. Executives must discuss and decide how to balance choices. Seeing the impact of the various options on the curve can help make the effect of the alternatives much more real.

Second, the curve can show the size of the risk more clearly—in the depth and width of the prelaunch investment and the length of the time to scale. Time is always a risk, and "longer" inherently means more risk. Competitors may make important moves, and the market may shift in ways that will affect payback.

Use of the curve makes the dimensions of the risk clearer and better able to be managed. Usually, the clarity that the curve brings makes companies more comfortable taking on appropriate risks, which allows them to shoot for more growth.

Conversely, an inadequate evaluation of the risk involved can lead a company to take on too much risk and lead to the creation of a cash trap.

FIGURE 2-5

Three types of risk

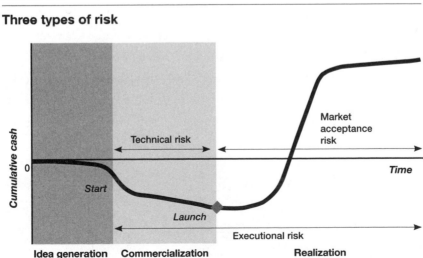

Cash Traps

More than thirty years ago, Bruce Henderson, founder of The Boston Consulting Group, wrote that "the majority of products in most companies are cash traps—they will absorb more forever than they will generate."[13] This is still true today. The stark reality (and often hidden truth) is that many new products and services, even seemingly successful ones, don't achieve payback over their lifetimes.

Today, in almost every industry, companies invest heavily in innovation activities, product life cycles continue to shrink, and copycat products erode pricing power with increasing rapidity, all of which make earning a return harder than ever. Cash traps never create net cash profits because much of what they generate has to be reinvested just to maintain their competitive position. Cash traps can debilitate a company and its portfolio.

Sometimes, a product or service that is a cash trap generates one or more of the indirect benefits of innovation that we described in the first chapter: acquisition of knowledge, enhancement to the brand, strengthening the ecosystem, and building the organization. Generally, however, investment in a cash trap to gain indirect benefits can lead to trouble.

The greatest danger of a cash trap comes when it isn't recognized, or is recognized but still supported, without achieving sufficient indirect benefits to warrant the expense. Jim O'Connor, corporate vice president and general manager of the Early Stage Accelerator at Motorola, describes the problem: "If the noncash benefits don't lead to cash sooner or later, they're a distraction. Some of the big failures I've seen come as a result of people expecting too much in some of those noncash areas that don't actually lead to cash."[14]

Although it seems obvious that companies would be on the lookout for cash traps, would be able to spot them relatively easily, and could get rid of them quickly, there is something in the nature of the cash trap, especially a large one, that makes it very difficult to kill. It may be that there is a bit of the gambler in every innovator, and they have a fundamental belief that every bet they make will pay off—someday.

Concorde: A Cash Trap

The Concorde supersonic aircraft, although an iconic invention and one of the most dramatic and well-publicized advances in aviation, was a cash trap. The Concorde had a futuristic look, with a needle-point droop-nose and swept-back delta-shaped wings. It was equipped with four afterburning Rolls-Royce/Snecma Olympus turbojets that were an evolution of a military design and it was the first commercial airliner that operated with fly-by-wire controls. With a cruising speed of Mach 2.02, the Concorde could zip across the Atlantic in three hours and forty-five minutes, which meant that it arrived in New York at an earlier hour than it had departed from London, thanks to the five-hour time difference.[15]

Companies in France and England separately began development of commercial supersonic engines in the early 1960s, but it soon became obvious that the start-up costs would be far too great for either company to bear alone. So France and the United Kingdom forged an alliance (negotiated as an international treaty, rather than as a commercial partnership), and joint development began in earnest in 1962.

The original estimate of the investment required for the Concorde was less than $1 billion, but it actually came in at about $4 billion (or about $11 billion in today's dollars). Spending four times the budgeted prelaunch investment would be enough to destroy the payback potential of almost any invention, but the Concorde also struggled through a much longer time to achieve scale than had been planned. It took fourteen years from first prototypes until British Airways (BA) and Air France (the launch customers) sold the first tickets. Even so, the Concorde's developers hoped they could recoup their investment by working with airlines to create a worldwide supersonic network of aircraft and flights catering to wealthy passengers who would have little or no price sensitivity. But even at a $9,300 round-trip fare between

London and New York—well above the first-class fare on a conventional trans-Atlantic flight—the Concorde needed to have very high utilization to achieve cash payback for a single flight, let alone recoup its entire investment. But, just as with the Iridium phone, the price proved prohibitive. Both Air France and British Airways often had flights with many empty seats and had to discount tickets or upgrade frequent fliers from regular flights to the Concorde to fill up flights.

What's more, the support costs required to keep the Concorde up to date had a negative effect on payback. The Olympus engine that powered the Concorde was based on models created in the 1950s, a time when fuel costs, noise, and environmental sensitivities were very different. Although the fly-by-wire systems were innovative in 1976, the controls and instrumentation were soon surpassed by conventional aircraft. After a fatal crash in 2000 (the only one the Concorde suffered), the aircraft were grounded. Although the Concorde briefly returned to service, it took its last flight in 2003.

For the government-owned (and government-funded) companies that created the Concorde, the development of the aircraft was almost certainly a cash trap. The huge original investment and the selling price to British Airways (BA) of a mere £1 each meant that the cash curve never rose above the cash breakeven line.[16] The airlines, BA and Air France (both of which were state owned for part of the time they were involved with the Concorde), operated the flights for almost three decades and, at best, achieved cash breakeven. Certainly, the companies and the governments that supported them believed that the indirect benefits—particularly to their corporate brands and national images—were significant.

iPod: A Model Cash Curve

For an example of superb cash curve management, take a quick look around you. Odds are that someone nearby is plugged into the payback phenomenon known as the iPod. Since its launch in late 2001,

Apple's digital music player has quickly become the single most successful consumer electronics product in history, eclipsing even the previous superstar, the Sony Walkman.

Most people know the iPod has been a smash hit. But what's behind Apple's success? Apple did not come up with the idea of a portable digital music player. In fact, three or four other companies (depending on how you count) had digital music players on the market before Apple did, including Diamond, which released the Rio, and even the computer company Compaq, which prototyped a hard drive player with many of the same specs as the first iPod.

Nor does the brilliance and appeal of Apple's design fully explain the continued success of different versions of the iPod. Indeed, Apple has had more than its fair share of failures, even ones with great design. In 2000, Steve Jobs hailed the company's Cube computer as "possibly the most stunning product we've ever made." After less than a year on the market, the Cube was shelved due to poor sales.

The real reason the iPod is so successful is that Apple has managed the cash curve expertly. It had a beautiful, almost perfect, curve—with a shallow dip of start-up costs, speedy journey to market, steep rise to scale, and high postlaunch profit (see figure 2-6).

We don't have enough inside information to draw a perfectly precise curve for the iPod (but we hope that Apple's CFO has done so), but by drawing on publicly available data and industry benchmarks, we can get close.

Start with the overall sales. Through the 2005 holidays, Apple sold 42 million iPod units in just over four years on the market.[17] All told, the various elements of the iPod business, including the iTunes Music Store, generated more than $7 billion in revenue for Apple from 2000 to 2004. That's where most analyses of the iPod end, but there is more to be learned by looking at the four Ss.

First, Apple managed to keep its start-up costs low. Over eight months of development, fewer than fifty people worked on the project team at any one time, in comparison to the scores they might deploy

FIGURE 2-6

iPod cash curve

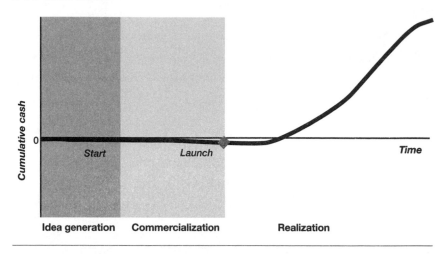

on the development of a new computer platform.[18] According to publicly available data and benchmarks from similar projects, we estimate that Apple spent around $10 million in 2001 developing the first iPod. Compare that with the $500 million reportedly spent on Apple's ill-fated Newton PDA before it was pulled from the market.

Second, Apple took the iPod to market with incredible speed—less than one year. In the spring of 2001 Steve Jobs asked Jon Rubenstein, Apple's senior vice president for hardware development, to take on the iPod project, and by November 2001 the product was on store shelves.

How did it achieve such speed? Usually, Apple prefers to create products based on its own technology. However, with the iPod, Apple realized that investing heavily in a custom design might not be the best choice.[19] Instead, not only did the company rely heavily on suppliers and partners for capabilities and expertise, it also used many off-the-shelf components.

Rather than assign a team of its engineers to design the "brains" of the iPod, Apple turned to PortalPlayer, a small company that already had developed a design for just such a device. Apple persuaded PortalPlayer to drop its other customers and focus exclusively on the iPod. Over the next few months, Apple worked closely with PortalPlayer and other key suppliers to flesh out the prototypes and overall design.

At the same time, Apple put considerable effort into developing relationships within the ecosystem, which enabled the company to create and launch the online iTunes Music Store. It became the first major successful realization of a total system of downloadable digital music that did not violate copyright laws and that benefited the content providers, Apple, and the consumer.

The first iPod products appeared at retail in November 2001, just in time for the holiday shopping season. Apple's marketing muscle kicked in—including the ubiquitous and much-praised advertising campaign featuring black-on-white silhouettes grooving and gyrating to their iPod sounds. Apple spent $28 million to advertise the launch in 2001, and a total of about $69 million between 2001 and 2003.[20] Sales rose sharply and the curve quickly arced upward toward optimal scale.

Apple also made a smart (and risky) move to control support costs and build postlaunch profit. During development, Apple had agreed with Toshiba Corporation, the world's only supplier of the mini hard drives used in the iPod, to buy its entire output for eighteen months.[21] Not only did Apple get a price break, it also prevented potential competitors from moving quickly to match the iPod capabilities. It was therefore able to concentrate on building the market rather than fighting to differentiate the iPod from me-too products. Capturing essentially all of the market, at a premium price, greatly enhanced cash payback.

Apple also did not allow itself to slow down its pace of innovation. Within four months, it introduced the next iteration of the iPod, the 10GB, and continued to rapidly expand the product line with a regular drumbeat of new additions, including a Windows version, a model capable of displaying photographs, two extremely small flash memory–based players—first the iPod shuffle, then the iPod nano—and a video-playing edition.

All these decisions combined to help the iPod's cash curve cross the cash breakeven line somewhere in 2004, and keep it rising steadily.

Cash in the Context of the Portfolio

Most companies maintain a portfolio of products and services, and each one has its own profile of returns. In a typical portfolio, there will be a small number of products that deliver high cash payback and a few that are cash traps; others will produce varying amounts of cash, often along with one or more of the indirect benefits.

For a single-product start-up company, of course, discussion about the overall portfolio of products and services is either abstract or ir-relevant, or both. For more established companies, however, the port-folio is essential. In a big corporation, everything is relative—whether an invention is worth pursuing, and how, is subject to a broad range of other factors and alternatives.

When the cash curve is not used to determine which new ideas to invest in and add to the portfolio, we find that the payback potential of the ideas is routinely overestimated. As a result, the portfolio fills up with new products and services that ultimately deliver less payback than expected and, in total, cannot generate enough return on the in-novation investment to enable the company to achieve its growth ob-jectives. When this happens, it may be necessary to significantly step up the company's innovation spending (with no confidence of success) or to explore other strategies for growth—including mergers, joint ven-tures, or alliances of various sorts.

However, when developing the portfolio, sometimes management will focus too much on the potential of each product for short-term cash payback and pay too little attention to indirect benefits that may lead to payback in the longer run. This lack of attention to knowledge acquisi-tion, brand building, ecosystem strengthening, or energizing of the or-ganization can cause nearly as much trouble as can too little cash.

Many portfolios become too heavily weighted toward only short-term cash payback, with little being developed for the future (in terms of either cash or indirect benefits). Often the resulting portfolios are lacking in new ideas that will have a payback in more than one fiscal year (or, alternatively, beyond the number of years the presiding executives expect to be in their positions). When this happens, companies may find themselves on a roller-coaster ride, with sharp spikes of payback followed by rapid descent into unprofitability.

Timing is an important issue in determining the interrelationship both of cash curves across different innovations and between cash payback and indirect benefits within the portfolio. All companies, regardless of the businesses they operate in, need to consider ideas that are likely to bring relatively immediate payback as well as ones that will generate payback at some later time. The percentage of a company's innovations that should be short term versus long term will vary by industry and by the company's competitive situation, but building a strong portfolio that delivers consistent cash payback, without neglecting the indirect paybacks, is one of the most important aspects of managing the innovation process.

THREE

The Indirect Benefits
of Innovation

Money is not, maybe, the whole answer.

—Claus Weyrich, Siemens

Yes, in addition to cash, there are other benefits to be gained from innovation.

When people are presented with the idea that cash payback must be the ultimate goal of innovation, they sometimes react with worry, alarm, and confusion. They worry that their organization may not be able to properly measure the cash payback from its innovation efforts. They're alarmed that once the cash payback can be and is measured, it won't look as rosy as it should. And they're confused because cash payback seems too limited a measure of innovation and doesn't allow for other ways of thinking about the value of innovation. What about creativity? Invention? Motivation? Making a positive contribution to the world?

Sometimes companies pursue innovation for reasons other than cash payback. They articulate those reasons in a variety of ways, such as "We'll learn a lot," or "It's good for morale," or "It will have a halo effect on the brand," or "It will make our partners happy." These goals may be, in fact, positive in their own right. But they are only legitimate reasons to innovate if they ultimately result in cash payback. Too often, however, they are cited as justifications for pursuing a questionable project—such as an executive's pet idea or a politically popular boondoggle—or as excuses for an effort that clearly has no chance of achieving payback.

There are four "indirect benefits" that can accrue to a company from its innovation activities:

- **Knowledge acquisition.** Innovation can add to a company's store of knowledge.

- **Brand enhancement.** Innovation can significantly strengthen a brand.

- **Ecosystem strength.** Innovation influences relationships with outsiders to the company.

- **Organizational vitality.** Innovation is often quite meaningful to people within the company.

To repeat, however: these are benefits—positive side effects—not paybacks in themselves. They are not truly valuable to the company unless and until they lead to cash payback. Their positive impact must be able to be seen on the cash curve of either current or future products or impact the performance of the company overall. Although the indirect benefits are not as easy to quantify as cash payback itself, their value is still measured by the cash they help return.

Knowledge Acquisition

The relationship between cash return and the acquisition of new knowledge is a complex one that lies at the heart of innovation. Although cash is the ultimate goal of innovation, the route to cash payback from innovation always leads through new knowledge. Without new knowledge, there can be no innovation.

The acquisition of new knowledge often requires a substantial investment of money and time. The cost can become so great that it can seriously reduce cash payback or even lead to the creation of a cash trap. It can take so much time as to double or triple time to market. A typical cause of delay occurs when a company tries to develop a new idea while simultaneously acquiring the new knowledge that is required to commercialize it. As the life cycles of new products shorten, companies are more and more tempted to bet "on the come"—forging ahead with development before the necessary knowledge is actually in hand, assuming that it can and will be acquired by the time it is needed.

When the knowledge needed is relatively limited or not particularly complex, the bet can often pay off. But as the technical complexity of the product increases, requiring many types of new knowledge in many parts of the product or process, the bet entails much greater risk. Recent high-profile delays in software, game consoles, automobiles, and many other products testify to the difficulty of trying to commercialize new products and acquire the required new knowledge at the same time.

"Many innovations came from research which started long ago," said Didier Roux, vice president of R&D at Saint-Gobain. "You can always tell the story in such a way that you believe at the end that it's recent research that makes the product. But when you look at a very innovative technology, you often find that the first research was started long, long before the innovation went on the market. Whatever you say and whatever people believe, some innovation takes a lot of time. If nobody does the research, then nobody will get breakthrough innovation on the market."[1]

The management challenge is to determine how much to invest in knowledge acquisition and when to do so. Sometimes companies do not invest enough, which can result in further delays, product shortcomings, or market indifference. Companies will also sometimes overinvest, when they convince themselves that a major cash payback can eventually be achieved. If the payback does not arrive as planned, the overinvestment may come back to haunt the company, by limiting its ability to invest in further knowledge acquisition (or other important areas).

The benefit of knowledge acquisition can go beyond an individual product or service and have an effect on the company's entire portfolio of offerings. Even when the acquisition of knowledge for a given product is highly costly and is certain to reduce the cash payback of that particular invention, the new knowledge may be extendible into other areas of activity within the company. It may be applied to improve current products and services that it was not originally intended to benefit. Or it may become the basis for currently undefined future products and services—and ultimately generate sufficient cash payback to make the investment worthwhile. In the end, to be valuable, all investments in knowledge must generate cash.

Four categories of knowledge can contribute to payback:

- **Product-specific knowledge.** Knowledge that will be used in the creation of a specific product.

- **Product-applicable knowledge.** Knowledge that does not have an immediate product application but that can be applied to the company's known categories or business areas.

- **Greenfield knowledge.** Knowledge acquired with the intention of opening whole new business areas or product categories.

- **Knowledge as product.** Knowledge managed as an asset in itself, through sale or license to other companies.

Product-Specific Knowledge

Knowledge can be acquired specifically and deliberately for the creation of a new product or service. Its contribution to payback is direct, and the timing of the payback can be planned.

When the new knowledge that will be required to create a new product or service is reasonably well defined, the cost of its acquisition and its effect on cash payback can be determined and managed without great difficulty. The investment required to gain the new knowledge can be considered a start-up cost, and its effect can clearly be seen on the start-up segment of the cash curve.

Consider, for example, a company that creates and markets kitchen knives. For a proposed new paring knife, it wants to develop a grip with a different material than it has used in any of its existing products. The material exists and has been incorporated into other categories of products, such as writing instruments, but the company needs knowledge about how the material will behave in this particular application, in combination with other materials, and in its manufacturing process.

Product-Applicable Knowledge

The process of acquiring new knowledge usually requires a certain amount of inquiry, investigation, and testing. These activities will then yield more information and understanding than is necessary for a specific product or service that has been proposed or is currently in development. Discoveries are made. Bits of interesting data are gathered. Promising avenues are opened.

Many companies, especially those with research capabilities, are constantly working on such knowledge and applying it to products and services that have been proposed, are in development, or are already in the portfolio and could be modified or improved. There is a reasonable certainty that the knowledge can be applied within a relatively short period of time, a year or two, and that it has the potential to contribute to payback.

For the knife company, product-applicable knowledge might involve research into how to improve the blade-sharpening process, with the goal of making knives sharper and more able to retain their edge longer under hard use. This knowledge will obviously be applicable to the company's current portfolio of products. However, there is no guarantee that such a process can be found, will be cost-effective, or will be accepted by the market.

The cost of acquiring product-applicable knowledge sometimes appears to be much less than it actually is. This is because it takes place over time and may not be considered as part of the start-up cost for a specific product (and is therefore often not well tracked). But the costs can become substantial. This is why companies expect their employees to manage the time and carefully track the amount of money they spend on quasi-independent projects for the development of product-applicable knowledge.

Greenfield Knowledge

Some companies, usually large and profitable ones, have the resources to invest in acquiring knowledge that they believe may have the potential to open whole new areas of business and as-yet-undetermined categories of products and services. This can be an expensive activity because it deals with multiple unknowns—technological, market related, and operational. More and more, companies that maintain basic research facilities are including people with business instincts and market knowledge in the process of greenfield knowledge development. The cost of acquiring such knowledge can grow so high that it can turn whole businesses into cash traps.

For example, suppose the knife company believes that lasers could be incorporated into a "bladeless" kitchen implement that enables people to slice foods with more precision than ever before and with no danger of slicing themselves. This would require greenfield knowledge. There is no guarantee that laser technology can be harnessed properly, that an implement can be made workable, can be manufactured, can be offered at a reasonable price, or will be accepted by the market.

Other companies will search even farther afield, in search of promising new knowledge that may yield opportunities completely unrelated to their current activities. Most of the companies investing in nanotechnology, for example, do not yet have a clear set of product plans to let them realize the necessary level of cash return on those investments, but are spending to develop knowledge to use across a number of yet-unknown opportunities.

Sony CSL: The New Knowledge Acquisition Lab

A ten-minute walk from the JR Gotanda station in Tokyo—along Sony Street, past the Cozy Corner store, where you can buy French-style pastries, and the 7-Eleven store, its shelves bursting with DVDs, fireworks, sushi, burritos, and frozen cocktails ("Gaijin love it!" proclaims a billboard)—brings a visitor to the Takanawa Muse building, where the Sony Computer Science Labs (CSL) are located.

Twice a year, the thirty researchers and scientists who work here hold a kind of Sony-internal trade fair of new knowledge and ideas. People from throughout the company are invited to visit the labs, view the work on display, talk with the scientists, and try to figure out how they might use the ideas in their own business areas. CSL was founded in 1988 to focus exclusively on issues directly involved with computer science, such as networks and programming languages. Since then, its work has broadened considerably to encompass related research into brain sciences, systems biology, and mechanisms of consciousness. "And the time frame for a result? Could be five to ten years," Mario Tokoro, the president of Sony CSL, told us.[2]

When we visited CSL during one of these internal exhibitions of new knowledge, we saw a demonstration of a new wireless technology that enabled transmission of content to very precise locations in a home, a music system that allowed friends with digital audio players to combine different songs into a blended composition as they moved toward each other, and a pair of robots whose sensing systems made it possible for them to find one another even in a crowded room, like an automated twenty-first-century Romeo and Juliet. All of these prototypes contained intriguing new knowledge, although none of them would necessarily become a product.

Sony's CSL is entirely devoted to the acquisition of new knowledge of all the types discussed earlier—product-specific, product-applicable, and greenfield. As Tokoro told us, one of the labs' main missions is to develop knowledge that can make "a contribution to current Sony products. Two examples are the touch-sensitive 'G-sense' panel that is used to control the VAIO Pocket HDD portable audio player. We also developed a PO Box (Predictive Operation Based on eXample), an input-supporting software for Japanese language in some of our VAIO models." CSL researchers are also engaged in developing new knowledge that has the potential of "opening up new businesses and new business areas," said Tokoro. Knowledge gained at CSL played an important role in the development of Sony's networking products, VAIO computers, and PlayStation.

As we see at CSL, new knowledge that is not product specific can improve the cash curves of future inventions. In particular, being able to leverage previously created knowledge can reduce the required level of start-up investment. Also, companies can avoid the problem of simultaneous development of both knowledge and product, which can improve their speed to market.

Maximizing Cash Return on Knowledge

Although there can be no innovation without new knowledge, it does not necessarily follow that new knowledge of the type acquired at Sony's CSL and other such facilities in companies around the world, will lead to cash payback for the company that developed it. In fact much new knowledge will quickly find its way into the broader industry and become commoditized, resulting in a weakening of the competitive advantage (and payback) the company had hoped to achieve.

That's why it is important for companies to protect the knowledge they acquire. Competitors will quickly take advantage of any unprotected new knowledge that comes into the market. As a result, the company's investment in new knowledge becomes, in effect, a form of industrial philanthropy.

In addition to the obvious (although often complicated) methods of protecting knowledge—patents and trademarks—there are many other ways to protect knowledge and, therefore, the cash payback that new knowledge can bring.

Limit availability. Target, the number-two discount chain in the United States, offers a variety of innovative products created by well-known designers—including Isaac Mizrahi (clothing), Amy Coe (children's bedding and accessories), Liz Lange (maternity), Mossimo (junior fashions), and Michael Graves (housewares)—that cannot be purchased at any other store. In effect, Target is making an investment in the knowledge of its design partners and betting that the investment will pay off because it will have control over where and how the products are realized.

Target's investment in developing and retailing limited-availability goods from creators with special market, product, and consumer knowledge has contributed to the company's strong payback. It is one of the few retailers that has been able to successfully compete with Wal-Mart, a retailer that is better known for process innovation than it is for product innovation.

Secure rare talent. Believing that certain people are particularly skilled at acquiring and developing new knowledge, many companies will make large investments in hiring, retaining, and nurturing them. Financial incentives are one favorite mechanism used to bind talent to a company, but there are many others, including job flexibility, ability to access resources, and public/peer recognition.

Microsoft famously used copious amounts of stock options to attract virtuoso programmers in its early days. Microsoft went public in 1989, not because it needed the capital (it enjoyed pretax margins of 34 percent at that time), but because a public offering was the best way to provide a liquid market for the stock it had granted its early employees.[3] This enabled the employees to sell some stock and realize cash gains, if they desired, and still stay with Microsoft. It also made the company more attractive for other talented people, who had many competing options, to join.

Dress it up. If an invention has a distinctively unique design or commercial image, it may be able to be legally protected by "trade dress" law. The specific capabilities and functionalities of a product can be protected by patents; trade dress offers legal protection for the nonfunctional attributes of a product. Callaway Golf's Big Bertha golf club, for instance, wasn't the first oversized golf club when it came on the market, but it contained several improvements in materials and composition that could be patented. The club also had a unique design that eliminated most of the weight from the club shaft and, most important, was instantly identifiable.

Callaway's founder, Ely Callaway, believed that the new knowledge contained in his Big Bertha club was so valuable that he secured both patents and trade dress protection for it. To further frustrate potential copycats, Callaway chose not to license the design or market the product through partners. As a result, Callaway Golf came to be regarded as an innovative company and built itself into a leading supplier of golf clubs, balls, and sportswear. And although plenty of imitators eventually came along, Callaway enjoyed a long window of highly profitable leadership.

Make it a trade secret. Pixar, the hugely successful producer of computer-animated feature films—including *Toy Story* (1995), *A Bug's Life* (1998), *Finding Nemo* (2003), and *The Incredibles* (2004)—developed the rendering system called RenderMan, which makes it possible to apply lifelike textures and lighting effects to objects (including actors) designed with other programs.

In 2000, four employees left Pixar to start Exluna, Inc., which developed a rendering program called Entropy.[4] Exluna priced its competitive system at about a third of the cost of RenderMan. In March 2002, Pixar sued Exluna, charging it with misappropriation of trade secrets, copyright infringement, and patent infringement. While RenderMan used an open standard, and many of its key algorithms were published in technical journals, Pixar argued that the key elements had been adapted and incorporated into the product in a unique way that qualified it for legal protection—not as a patent, but as a trade secret. In July 2002, the two companies reached an out-of-court settlement, and Exluna agreed not to sell Entropy to any new customers.[5]

Knowledge as a Product

Even if new knowledge cannot be applied to any of a company's current products, and seems unlikely to contribute to any future products or new business endeavors, it can still be valuable and be leveraged to generate a cash return—through sale or licensing to others. In effect, knowledge becomes a product in and of itself. There is a rapidly growing demand for knowledge as a source of competitive advantage throughout all industries and a sophisticated worldwide infrastructure for marketing, trading, protecting, and receiving value from intellectual assets.

Partially as a result of this burgeoning market for knowledge, it has become increasingly important for companies to know not only how to develop knowledge internally that will create competitive advantage for themselves, but also how to source knowledge externally, sell their store of knowledge (to others who want to use it to build up their own positions) and how to protect themselves at the same time. (We will discuss this route to cash payback in chapter 6, The Licensor.)

Brand Enhancement

An association with innovation can enhance the reputation of a company and its brand.

However, just as with knowledge, for a brand benefit to be valuable it must ultimately deliver cash payback, in at least one of three ways:

- **Premium prices.** Companies that are perceived as innovative often can charge higher prices for their products and services than their competitors can. The increased amount of cash generated enables the curve to cross the payback line more quickly.

- **Higher volumes.** Whether they are first to the market or not, companies that have brands that stand for being innovative usually have a substantial leg up on competitors, even those that are earlier to market. This can reduce the time it takes to achieve scale.

- **Greater acceptance.** It is often easier for an innovative brand to move into new product and service areas than brands that are seen as being conventional. Customers expect an innovative brand to explore new avenues for future growth and are more willing to follow them into uncharted territory. This acceptance helps the new product achieve scale quickly. It can also reduce support costs, because less persuasion and education may be required to get customers to buy.

Premium Prices and Higher Volumes: How Samsung Built an Innovative Brand

A decade ago, consumers thought of the Samsung brand, if they thought of it at all, as synonymous with cheap, low-quality, copycat electronics and appliances. In 2005, Samsung was named the number-one consumer electronics brand in the world in the *BusinessWeek/* Interbrand survey of the top 100 global brands. Samsung built its brand by focusing on innovation in the digital convergence space, which many other companies were trying to do, but where few succeeded as well as Samsung.

The company's success has its roots in the "new management philosophy" first articulated by chairman Kun-Hee Lee in 1993 and then reintroduced in 1997, following the Asian financial crisis. The philosophy has a number of key tenets, four of which directly affect the brand:

- **Quality over quantity.** On one of our visits to Samsung in Korea, we were told about a now-famous incident that took place in 1995, when the chairman learned of defects in mobile phones he had sent out as New Year's gifts. He ordered that $50 million of the faulty electronics components be piled in a factory courtyard, where employees smashed and burned them. Lee and the company's directors sat beneath a banner reading, "Quality Is My Pride," and watched in approval.[6]

- **Innovation.** Samsung underscored its commitment to innovation by doubling its spending on R&D. Over the years, the company has steadily increased its R&D budget and now invests about $5 billion each year. Few, if any, other large technology companies spend as high a percentage of their revenue on R&D.

- **Market leader.** Samsung committed to becoming the leader in each of its markets. As a result, Samsung was among the first, and most aggressive, of the big electronics firms to move from analog technology to digital. "The [digital] market is here now, but in 1997–98, it was a very unusual move," recalled Mr. Lim Sun Hong, vice president, Brand Strategy, Global Marketing Operations. "It was almost a futuristic concept."[7]

- **Increasing value.** Samsung quickly and aggressively began to exit businesses and discontinue products that were not growing in profit or perceived as innovative and thus did not enhance the brand. "Last year," Lim said, "we dropped the VCR business even though it was making money." While making VCRs is no longer anybody's idea of innovation, few companies are willing to actually stop selling profitable products to help their brand. Samsung was.

    As the Samsung brand gained luster, the company could command a price premium for its products and sell at high volumes at the same time. In 2003, Samsung outsold Sony in the North American projection TV market for the first time, despite Samsung's significant price premium.[8] Samsung even enjoys a slight price premium for its memory chips, products usually seen as true commodities.[9]

As a result, Samsung is now ranked the twentieth most valuable brand overall, eight places ahead of Sony. Over the past five years, Samsung's increase in brand value has outstripped every other global brand, enabling it to enter lucrative new markets, including China. "In China," Lim said, "young people try to buy Samsung products, even if their income isn't very high. Samsung has become a status symbol."

Greater Acceptance: Moving into New Areas

W. L. Gore & Associates, a company recognized as being highly innovative, has successfully ventured into areas outside its original business, helped greatly by the strength of its brand name.

Gore's best-known product is Gore-Tex, a waterproof and breathable fabric that is used in a wide variety of apparel, particularly outerwear. Over the years since the company's founding in 1958, Gore has applied its technology to products and technologies in a wide variety of product categories, including electronic signal transmission, fabric laminates, medical implants, and dental floss, as well as membrane, filtration, sealant, and fiber technologies for diverse industries—the strength of the brand name helping them gain early trial and fast acceptance in areas where Gore previously had little or no presence.

Brand Transformation Is Not for the Faint of Heart

Like the knowledge benefit, the benefit to a brand from innovation often will help the cash curve of many different products in the company's portfolio. But investing in innovation with the deliberate purpose of building or changing the positioning of a brand is not for the faint of heart. Brands in which innovation is important, whether in the business-to-consumer or business-to-business arena, are built on performance. The brand perception follows the results, not vice versa. Sometimes a company that wants to be seen as innovative proclaims that it already is, before achieving any results that would support the contention. Such proclamations only make management look foolish and do damage to the brand.

Ecosystem Strength

No company operates autonomously, especially in the age of global competition, connectivity, and worldwide markets. Each company is part of an ecosystem, a network of different organizations and entities (and sometimes specific individuals or the public at large) where the participants may be dependent on, support, or exist in a symbiotic relationship with others in the network. As such, companies must be attuned to the needs and desires of the other members of their ecosystem, since the benefits that strong relationships can bring are large and considerable pitfalls await if those relationships are mismanaged.

It is quite common that a company shares part of its ecosystem with its competitors. Distribution channels and suppliers are examples. These participants all have elements that are valuable but are in limited supply—management attention, training slots, shelf space, and others. Competition between companies, therefore, takes place not only at the end-customer level (where a product or service is sold) but also within ecosystems—to get what you need, when you need it. And that competition is often a zero-sum game. There is only so much of anything to go around, and if you get it, whatever "it" is, your competitor can't—and vice versa.

Some companies successfully use innovation to strengthen their ecosystem, often in ways that would be difficult or prohibitively expensive to do via other avenues. Innovation can help strengthen ecosystem relationships and improve cash payback in three ways:

- **Preference.** A company that focuses on innovation may be given preference over its competitors by ecosystem partners.

- **Exclusivity.** The innovative company may be able to develop exclusive ecosystem relationships.

- **Standards.** The innovative company may be able to gain support for an industry standard it favors.

Preference

Innovation can directly help strengthen an ecosystem by providing the innovating company with preferential access. This can take many forms—for example, more management attention or increased access to the sales force that will actually be dealing with customers. The engineering organization of an original equipment manufacturer (OEM) may give preference to a company that is perceived as the most innovative of the OEM's suppliers. The engineering group will do this because they believe that the innovative supplier will be able to provide the most advanced assistance, when it is needed, and thus reduce the amount of work the OEM must do.

In these cases, innovation delivers an important ecosystem benefit. Of course, often the same benefits can be achieved the good old-fashioned way—by paying for them. And that is exactly what an innovative company's competitors often find themselves forced to do, when the innovator gains preference. The impact of preference on the cash curve can be significant and multifaceted. If, for example, you are a car company that is seen as innovative, suppliers will want to work with you on their newest technologies. This can reduce your start-up costs, because suppliers are often willing to shoulder more of the expense in order to get their product on one of your car platforms. And in many industries, if the sales channel sees you as innovative, your support costs will be lower, and you can achieve scale faster.

Linde, the material-handling equipment maker, decided to invest in the development of a new drive axle for its forklift trucks, almost solely to achieve an ecosystem benefit. The start-up costs would be high, and Linde knew that it could not charge a premium for the new feature, because end users of the forklifts would notice only a small difference in performance.

However, Linde considered it a good investment for another reason—the improvements to the axle would make a big difference to the dealers who sold the forklifts to the end users. Linde knew its dealers very well and understood that a significant percentage of the forklifts were on long-term rental contracts to the users. This meant that the forklifts were still owned by the dealers, who paid for service themselves and who, obviously, cared greatly about how long the parts would last and how much labor was required to repair the vehicle. The improvements in the axle helped increase the service intervals, which had been as short as every five hundred hours of use, to as long as three thousand hours.

By reducing the dealers' costs of maintenance and repair, Linde directly boosted its channels' margins and increased its own profits. Linde has thereby strengthened its relationships with dealers who are now more likely to buy Linde machines and recommend them to their customers. This dealer preference is extremely important because Linde products are more complex than those of their competitors and are priced higher, which mean they require more dealer support and attention to sell. As Stefan Rinck at Linde put it to us, "This is something where we do something for our dealers to improve their profit margin. And at the end of the day, it's good for us to have dealers that enjoy good margins and are happy. If they are, they are willing to invest in training for their salespeople and other requirements needed to effectively sell our new products."[10]

Exclusivity

Preference in an ecosystem is desirable, but exclusivity is far more powerful. Many companies attempt to use their invention capabilities as a way to secure exclusive access to either a distribution channel or a customer group, or both.

For example, Whirlpool has for many years been the exclusive supplier of laundry products (washers and dryers) to Sears under the Kenmore brand. Sears is the largest seller of appliances in America, with the best-trained sales force, and for many years consistently grew its share at the expense of other industry participants. The Kenmore brand, which is owned by Sears, is the largest appliance brand in America.[11]

Because of a number of factors, of which its innovation capabilities is a very important one, Whirlpool has been able to maintain a position as the exclusive supplier of these products to this important customer and brand. Many other companies have attempted to get a piece of the Sears business, but Whirlpool's ability to consistently bring industry-leading innovation to the Kenmore brand, along with attractive commercial terms, has allowed it to retain its exclusive position with Sears. This has been a win for Sears, because the company gets substantial innovation into its Kenmore brand without the expense of its own research and development activities. Whirlpool has also benefited because it has been able to secure steady and substantial sales. Because of Sears's volumes, Whirlpool is able to much more quickly get to scale for its new products and services. This makes for an attractive cash curve for these new items, and the ability to achieve this scale also helps it build and position the Whirlpool brand in the minds of consumers. Also, Sears's extensive sales support for Kenmore helps reduce Whirlpool's need to invest its own money in support costs.

Standards

Strong ecosystem relationships can help an innovator create and maintain standards, which can be a powerful contributor to cash payback, both through increased product sales and via the licensing of the knowledge created.

Sony and Toshiba, for example, both have developed technologies that could be the basis of the standard for the next generation of optical disk systems. Sony's Blu-ray Disc technology offers more storage space than Toshiba's, but Toshiba's will likely be easier and less costly to manufacture. Both companies have been working to convince others in their ecosystem—including hardware manufacturers, computer makers, content suppliers, and retail players—to support their technology.

The process of gaining support goes beyond commercial terms and salesmanship, and affects the design and development of the technology itself. Sony, for example, has incorporated technology into Blu-ray Disc that prevents content from being illegally copied from a disk— "ripped"—and stored on a personal computer. Content providers, such as movie and television studios, liked this feature enough to throw their support to Blu-ray Disc. Sony also announced that it would incorporate the Blu-ray Disc technology into its next-generation video game system, PlayStation 3, which will broaden the number of Blu-ray Disc-equipped machines. As a result of these and other moves, Sony has created a large roster of supporters for Blu-ray Disc, including six of the seven major motion picture studios, as well as Philips, Hitachi, Panasonic, Sharp, Mitsubishi, Samsung, Pioneer, JVC, Dell, Hewlett-Packard (HP), and Apple.[12]

In addition to the cash curve of the new products Sony and Toshiba will sell, the company that creates the standard can gain further payback through licensing it. The standard becomes another business that has low support costs and can deliver exceptional return. This second cash curve, from the licensing of the standard, is what makes the ability to create standards so attractive.

The Risks of Innovating for Ecosystem Strength

The challenge with innovating to generate an ecosystem payback is one of balance. It is easy to help other ecosystem members make a great deal of money, but at an unacceptably high cost to yourself. Undertaking innovations with the goal of generating an ecosystem payback needs to be carefully thought through and leveraged into preference, exclusivity, or a standard that leads to a cash payback. It is also important to make sure that the ecosystem is fully aware of the activities you are pursuing on their behalf, so you can get paid for them.

Organizational Vitality

Sometimes the underlying purpose of a particular invention, or even a program of innovation, is to benefit the organization that is developing the new product or service. In this situation, the resulting cash payback must come not from the specific invention or program itself but from the broader and longer-term impact it will have on the organization. The caution, of course, is to determine exactly how the organizational benefit will affect a company's ability to generate payback. We have often seen companies embark on the development of a new product or service in order to bring about an organizational benefit without much thought about how it will demonstrably improve the company's ability to create payback.

Two main organizational benefits can be gained through innovation:

- **Confidence.** Innovation can make an organization believe in its abilities to achieve payback, and as a result make it more willing to pursue attractive, but risky, opportunities.

- **Attractiveness.** People with innovative ideas and perspectives usually want to work in innovative companies.

Confidence

Organizational benefits are real, if elusive. When an organization believes it is innovative, and has confidence that it can develop something new that people want and will be amazed by (and willing to pay for), it is often ready and able to identify and attempt things that less confident organizations shy away from. When people feel energized and powerful, they develop new products and services with cash curves that have higher risk, and higher potential for payback. And by doing so repeatedly, they build their confidence and capabilities even further, so they are willing to go through the innovation process again and again. Over time, the advantage gained from the learning and increased number of attempts enables the company to innovate faster, more consistently, and more successfully.

Attractiveness

Innovation can also help an organization be a place that can more easily attract, motivate, and retain talent.

People are not wealth-maximizing machines. While cash payback may be the ultimate measure of "winning" for many companies (at least in the eyes of shareholders), this is generally not the case for most individuals. People want to be part of a company that contributes to some sort of greater good and can have an impact on society—and to feel that they are playing a useful role. For many, being part of an innovative company fulfills this need.

Bombardier Transportation, a leading maker of trains and other products, had not been known as an innovative company for many years but is now in the process of transforming itself. Ake Wennberg, vice president and chief technical officer at Bombardier Transportation, told us, "Innovation is very important for us internally, for our own people. We want to be seen by our internal human resources as a company for young people, one with an innovative image. A lot of people were seeing us as very old fashioned, very conventional, but there's a tremendous amount of development here. We need to have those innovative people. We started to talk about how we could become more innovative. How can we make sure that we really generate ideas? How can we put the process in place that encourages people to come up with innovative ideas? We did this initially for internal purposes. Of course, we also want to see, in the end, the economic benefit of it. But it started as a discussion—how can we be perceived internally as an innovative company?"[13]

The managerial challenge with the organizational benefits of innovation is knowing when and to what extent to undertake an effort aimed primarily at achieving those benefits rather than maximizing cash.

Innovation Projects Can Build Confidence

The idea of developing a superbly innovative product line to inspire the organization—even if it may not produce much cash payback—is used by companies in many industries throughout the world.

ZF Sachs AG, the German producer of clutches, shock absorbers, and other auto parts, invests a significant amount of time and energy in products for race cars, including participating in the Formula One circuit. These cars race at speeds in excess of 180 miles per hour and rely on highly advanced aerodynamics, electronics, and mechanical components. ZF doesn't participate in Formula One racing only for the cash payback—it does it also for the organizational benefit, as well as the knowledge it can acquire.

"It's a big deal for us, internally, for motivation," Peter Ottenbruch, a member of the management board at ZF Sachs, said. "We hope that what we do with Formula One will help people understand we have a culture that allows motivation. We want to make sure that we are seen as a company that deals in interesting things and that might be a good place for young engineers or technicians to work. It's also helpful to get engineers into the Formula One group, because it's a totally different approach to development. Projects in Formula One are counted in hours, not weeks or months. So it's the process and the approaches to problems that are interesting, because the challenges force people to think in a different way and hopefully begin to imagine how things could be totally different."[14]

The use of special innovation projects designed to inspire people and focus the organization is not limited to high-technology companies.

In 2002, The Boston Beer Company, home of Samuel Adams Boston Lager, launched a beer called Samuel Adams Utopias. It is a brew with the highest concentration of alcohol in the world, an almost unbelievable 25 percent, in comparison to the 5 or 6 percent of a typical beer. Although Utopias was costly to create and is unlikely ever to sell in large volumes, Boston Beer founder Jim Koch decided to develop the beer to achieve a number of indirect benefits—knowledge acquisition, brand enhancement, and energizing of the organization.

The journey to Utopias began in 1993 with the development of Samuel Adams Triple Bock, a beer that contained 17 percent alcohol. Boston Beer brewed the beer and then sent it to a winery to be aged. Koch experimented with various types of barrels, first with ones originally used to age bourbon, and then with sherry and cognac barrels imported from Europe. After aging, the Triple Bock was shipped to a bottling facility, where it was poured into cobalt blue–colored glass bottles (inscribed with 24-carat gold leaf lettering) and stopped with a sherry-style cork. The Triple Bock retailed for $100 a case, four times the price of conventional beer, and sold briskly.

The success of the Triple Bock made it clear that an innovative product could break through any price ceiling, and paved the way for the development of an even stronger beer. Utopias is aged up to fourteen years and retails for about $100 a bottle. Since its launch, Boston Beer has sold all of the eight thousand bottles it produces every year. While the product may at best be marginally profitable, Koch says the benefit comes in the pride the company derives from creating a one-of-a-kind beer and creating a whole new category of high-alcohol, long-aged "extreme" beers.

A side effect is that the success of Utopias has also helped institutionalize innovation at the company. Employees are expected to bring out a new extreme beer every year (such as Chocolate Bock) to earn their annual bonus. In addition, Koch says that extreme beers have fueled better collaboration within the organization since brewmasters and marketers have to work together to develop successful new products.

Translating Organizational Benefit into Cash

As ZF and Boston Beer have learned, inventions pursued for the indirect organizational benefits may or may not also have an attractive cash curve. But the organizational confidence, morale, and energy gained in a single innovation effort can help the cash curves of other activities in a number of ways.

For example, the organization may become faster to act, with less hesitation and indecision. It may become more efficient, with less effort and fewer resources wasted on indecision and false starts. It may become less fearful and more willing to challenge existing constraints. It can become more knowledgeable and skilled at innovation. And it is usually able to generate better ideas to start the whole process. All these traits can reduce the amount of start-up cost required for the creation of a new product or service, speed up its time to market, reduce its time to scale, and lower its support costs. All of which improve cash payback.

Organizational benefits can also be translated into cash by avoiding other expenditures. The cost of attracting and retaining talent is on the rise, as is the cost of replacing valued employees who leave. One of the major issues facing many large companies today is competition for their employees from other companies (often smaller ones) that are seen as more innovative. Witness the large companies in the computer, software, and online industries that have to fight daily battles to keep their employees, even when the condition of the stock market no longer makes start-ups seem like guaranteed winning lottery tickets. A track record and reputation for being innovative can be of significant value in attracting employees.

In addition, companies that have strong reputations for being innovative don't necessarily need to offer salaries at the top of the pay scale for their industry. Very few highly innovative companies are also known for industry-leading salaries and benefits. This, too, helps reduce the companies' costs.

Building confidence, increasing attractiveness, gaining speed, and reducing cost are all important aspects of the organizational benefit of innovation—because they can lead to payback.

Achieving a Balance of Cash and Indirect Benefits

When a company is considering the potential indirect benefits that may result from the creation of a new product or service, and how and when they may contribute to cash payback, there is no formula to follow. It is a matter of determining what the company and its portfolio needs most at the moment and are likely to need in the future. In other words, it's a matter of managerial judgment.

Evaluating the impact that any individual invention may have on the company and its portfolio is rarely straightforward. The only time the payback decision is clear, when you know for sure what matters and how to make the required trade-offs among the paybacks, is when a company requires cash to the exclusion of everything else to ensure its continued operation or survival.

If the company can invest in "noncash" projects, it's important to be as sure as possible that the new product or service really will create an indirect benefit, of the type expected, and that it will ultimately prove to be valuable. This involves some instinct and intuition, but it also requires some careful thought about how that benefit will lead to cash payback. Will an improvement in morale enable the company to tackle some other major current challenge more successfully? Will an enhancement of the brand be important in the launch of a breakthrough product that is currently in development? Too often, an indirect benefit sounds good but ends up having no impact on the company's ability to gain payback.

Decisions about innovating for indirect benefits also need to be made in the context of the entire portfolio of current products and services and those in development. It's important to analyze the cash potential of the entire portfolio and then think about the contribution of each project to immediate cash and future cash via the indirect benefits. This means:

- *Clarifying objectives* for each project, relative to indirect benefits.

- Carefully *tracking the investment* in projects that are not about direct cash.

- *Assessing results*. How has the project contributed to the company's ability to innovate, both qualitatively and quantitatively?

Although companies will often pursue indirect benefits without enough attention to their ability to generate cash, companies will also build portfolios of products and services that do not deliver enough indirect benefit. This lack of attention to knowledge acquisition, brand building, ecosystem strengthening, or vitalizing of the organization can cause as much trouble as too little cash.

One of the reasons for this lack of focus on the indirect benefits is that when an idea is chosen primarily for its ability to create a noncash benefit, it usually requires a champion with enough influence, authority, or persuasiveness to get the organization to accept the idea. People with the required conviction and power are scarce.

Jim O'Connor at Motorola described the kind of argument that a senior executive must make to push along an idea that is likely to deliver more indirect benefit than it will create direct cash payback. The champion "will have to say something like, 'The new idea is not going to get us a good financial return right now. I can't show the payoff for a few years, if at all, but I'm willing to take that financial risk. There are things we need to learn and know. If we don't put some money into this now, we might be out of the game altogether.'"

PART TWO

Choosing the Optimal Model

THERE ARE THREE innovation business models: integration, orchestration, and licensing. The choice of model can have a dramatic effect on a company's ability to successfully achieve payback with a new product or service, and also determines how payback, the indirect benefits, and risk are allocated between different parties involved in the innovation process.

There is no single "right" or "best" innovation business model. All will lead to payback when used in the appropriate situation. Most large companies will use all three models at once, depending on the project and the conditions within the company and the market. The choice of model should be made explicitly rather than by default, reviewed regularly, and changed when a different model may improve payback.

In chapter 4, The Integrator, we look at companies that choose to "do it all themselves" in order to retain maximum control over every aspect of the innovation process and to keep the lion's share of the payback. Because integrators shoulder most of the investment, they also bear most of the risk.

In chapter 5, The Orchestrator, we examine the popular model of innovation collaboration. Orchestration is not outsourcing and requires a very different set of skills than does integration—primarily in managing a network of relationships. Orchestrators share risk and payback with their partners but face dangers that integrators do not, including the possibility that a collaborator will become a competitor.

In chapter 6, The Licensor, we show how more and more companies are choosing to achieve payback on their ideas and intellectual assets without having to make an investment in commercialization or realization.

FOUR

The Integrator

The engine is the holy shrine.

—Martin Ertl, leader of Innovation Impulses, BMW

THEY CALL IT THE FIZ, the glass and polished-metal facility just outside Munich where some seven thousand engineers, prototype builders, computer experts, and scientists across multiple disciplines work together in a huge space to invent BMW's future.

When we visited the FIZ (an acronym for the German *Forschungs- und Innovationszentrum*, also known as the BMW Group Research and Innovation Center), we entered the soaring, sun-strewn reception area, where we were required to leave our photo ID, and were then directed to the floor that serves as the home for ES20—BMW's Innovation Management Department. We went into the only space on the floor that is enclosed with walls—a small conference room tucked away in a corner. Within our view, members of the department—along with members of the seven innovation councils composed of people from the specialist KIFA departments (body, electronics, chassis, and drive train)—were focusing on various technologies and systems that might one day become part of a BMW vehicle.

Although an automobile is an invention of many parts, a BMW is all about driving pleasure, and the heartbeat of that is the engine, which is why the company follows the integrator model to create its engines. It is not only a matter of control and protection of intellectual assets, it is far deeper and more essential than that: designing, developing, and manufacturing the engine is the primary mission that defines and distinguishes BMW from every other car company and, for that matter, every other enterprise on earth.

BMW manages virtually every aspect of creating the engine, from forging the metal to manufacturing almost every part and assembling the final product. How could an activity that is so central to the company be created in any other way? As Martin Ertl, leader of Innovation Impulses at BMW, said, "As a manufacturer, you have to really focus on your strategic core areas. And for BMW that is engine development. We're willing to work with suppliers on anything else, but not design and not the engine."[1]

BMW could follow a different path. It could orchestrate the creation of its engines by managing several innovation partners and taking the role of "final authority" on exactly how all the discrete elements come together. But the company has deliberately chosen not to do so, and has aligned itself around this decision. The results have been dramatic. Not only is BMW seen as one of the world's most innovative companies (number sixteen on our 2006 BCG/*BusinessWeek* "Senior Management Survey on Innovation"), it has achieved remarkable cash payback. In 2005, BMW had among the highest operating margins of any automotive company in the world.

This does not mean that integration is the right business innovation model for every automaker, or any specific company for that matter. It means that BMW has executed the integration model with exceptional skill to achieve an admirable return. Not only did it create an attractive cash payback, it also generated significant indirect benefits and effectively leveraged them for payback.

Although the cash return is easily seen in BMW's remarkable profitability, several other cash curve dynamics are also noteworthy. Due to in-house integration of development and design, there is larger prelaunch investment for BMW's products, represented by a deeper dip in the start-up portion of the curve. However, because of BMW's core capability in engine development, the time to volume may be shorter, therefore reducing the time for the cash curve to cross the cash breakeven line into cash-positive territory. Also, customers are willing to pay a premium for vehicles designed and built by BMW, which affects both the rise of the curve and its eventual height. The shape of the BMW cash curve looks significantly different from the curve of a car from most other automakers.

Integration also brings indirect benefits to BMW. The knowledge it develops through its research is tightly aligned to its longer-term product strategies, ensuring that it gets used (and turned into cash). Its stream of engine innovation supports its brand as "the ultimate driving machine," and the resulting price premium. BMW is the partner of choice in its ecosystem due to its reputation for innovation and success in bringing new advances to market. Finally, the best automotive engineering talent also wants to work for BMW; it was voted the most ideal employer by engineering and science students in Europe in 2005.[2]

Intel Corporation, too, is an integrator whose main focus is the microprocessor, and has also achieved extraordinary payback through the do-it-all innovation model. Intel invests more than $4 billion a year in R&D and has approximately seven thousand researchers around the world pushing for the next advances in semiconductors as well as advanced computing, communications, and wireless technologies. Intel also operates fifteen major manufacturing facilities worldwide, and almost half its employees work in the manufacturing group.[3] All of Intel's plants are almost exact copies of each other, from the way they operate down to the smallest details.

For Intel, the combination of world-class R&D, massive scale, and extreme standardization is all about control. The company believes that if it controls literally every aspect of idea generation, commercialization, and realization, it controls its destiny. Its aggressive R&D investments are designed to keep it at the forefront of technological advances. Simultaneously, the company works hard to push the envelope as one of the world's largest and most sophisticated manufacturers, spending more than $3 billion a year on capital improvements. Intel's integrator model creates a cash curve with a deep prelaunch investment "hole," which is offset by the shorter time to market and time to volume that come with the standardized approach and focus on executional excellence. Intel also invests heavily in realization and, of course, makes a substantial postlaunch investment in marketing activities, including the "Intel Inside" and "Leap Ahead" advertising campaigns.

Intel's microprocessors and BMW's engines are quintessential examples of the integration model.

The Payback and Indirect Benefits of Integration

Companies choose to integrate for many reasons. The integrator is the sole owner and executor of the innovation, and it is the primary, if not the only, participant in the rewards.

Companies usually choose to integrate because they believe they are best able to manage the commercialization and realization of an idea (whether it is their own or comes from outside) by themselves, and because they don't wish to take on the risks of working with others—risks that include being slowed down, having their ideas stolen, or having value siphoned away.

In addition, companies choose integration because they believe it will give them greater control over costs and timing, since they will be

completely responsible for managing their own operations and investments. And as the company moves along the experience curve, it will
begin to operate more efficiently than if it were reliant on other companies. In theory, at least, for any new product or service, the potential cash curve for an integrator frequently appears more attractive
than it does for either of the other approaches. Since the company has
made the investment itself and is not sharing margins with partners,
the cash payback—if the new product or service is successful—stands
to be proportionately larger as well.

But the integrator also takes on the lion's share of the financial risk
by shouldering most, if not all, of the investment. Generally speaking,
the integrator will incur greater up-front costs than an orchestrator or a
licensor would. For instance, if the acquisition of greenfield knowledge
is involved, that doesn't come cheap. Nor does manufacturing, if that
has to be part of the equation. If the company has global ambitions, it's
no insignificant task to invest in the infrastructure and additional resources required to establish and maintain a distinctive, compelling,
and consistent presence in multiple markets worldwide. The integrator
exposes itself to a significantly greater risk of suffering a major cash
loss, but also sets itself up for a potentially large cash payback.

Successful integration, therefore, can be expensive and complex,
which, indeed, is one of the reasons why companies are increasingly
moving to one or both of the other innovation business models. In terms
of the cash curve, a move from integration to orchestration means
that the company makes a trade-off. Instead of the relative certainty
of a large cash investment and an uncertain but potentially rapid and
steep rise to scale and payback, the company opts for a shallower
start-up hole and a greater ability to reduce support costs if the curve
fails to rise as planned.

Indirect Benefits

In addition to the potentially higher cash payback, integration can
deliver significant indirect benefits. Handling all aspects of development

can provide the company with a unique and comprehensive body of knowledge. As an integrator, the company is not expert at one facet; it needs to be expert at all. Insights into one area can directly inform others. In addition, more and deeper knowledge of the full scope of activities can translate into improvements in operational performance and greater leverage of innovation spending.

For the brand, many customers still feel more trust in a company that has control of the entire process—there may be an assumption that quality is higher. More important, if an integrator is able to build the capabilities required to consistently produce distinctive new offerings, the brand may improve as innovation becomes something the company is actually delivering and known for. Similarly, for the people both in the integrator's ecosystem and within its own organization, integrating may tie them closer to the company.

Companies of All Types Can Succeed with Integration

People often think of the traditional auto manufacturer as the quintessential integrator, the company that owns the innovation process from first pencil sketch of the new model to its delivery to the dealer. But many other types of companies, large and small, in a wide range of industries, employ the integrator innovation business model.

Del Monte Fresh Produce Company, for instance, calls itself "one of the world's leading vertically integrated producers, marketers, and distributors of fresh and fresh-cut fruit and vegetables, and a top producer, marketer, and distributor of prepared fruit, vegetables, and other products in Europe, Africa, and the Middle East," according to company spokesperson Christine Cannella. The company, founded in 1886, owns its own plantations, growing its famous "Gold" pineapples and other produce, and employs its own scientists to study new plant varieties and processing improvements. Del Monte owns or charters refrigerated vessels that transport its produce. It also owns or leases the packaging centers, distribution centers, and even the trucks that bring its products to market.

Del Monte says its integrator approach from the field to the kitchen allows it to develop unique products, control the quality of its produce, and respond quickly to new demands or trends. Whatever the integration model may add in cost, Del Monte believes it is more than made up for by the incremental revenue generated by the high quality and responsiveness it achieves. It's an explicit trade-off made by the company—increasing cash expenditures to deliver a better and differentiated product to the market. Having a better, more distinct product to offer helps drive revenue and fuel top-line growth.

Of course, to say that the integrator owns and does everything is a bit of an exaggeration; all integrators work with suppliers, vendors, contractors, advisers, and many other collaborators and partners. Still, the integrator maintains a level of control that companies using other models do not. In addition, the integrator holds the major responsibility for actually creating the product or service, or at least the most essential elements—the ones that, if compromised in any way, could put the entire innovation in jeopardy and even threaten the success of the company itself.

Integrators are not necessarily big companies, nor do they necessarily have big R&D capabilities. The ability to conduct research and development, especially into greenfield areas, is not the defining characteristic of an integrator. Consider financial services, for instance, and integrators like Citibank and HSBC: their ideas and inventions spring primarily from their experiences in a wide range of markets—serving many different segments with many different financial products—not from blue-sky research centers. While their start-up costs to innovate are not as large as those for the manufacturing companies we have discussed, they still focus on the same cash curve levers by identifying new segments to drive up payback.

ECCO: A Little Integrator

ECCO, the Danish shoemaker, is a fraction of the size of big integrators like Intel and Samsung, yet ECCO's strategy and approach to innovation are remarkably similar to theirs. While many shoe and apparel companies have long outsourced production, ECCO believes that keeping a competitive edge requires achieving and maintaining mastery of key technologies—in particular, the unique direct-injection method that enables ECCO to create shoes of high quality and superior comfort.

"In a competitive market, we create results through our active and deliberate choices," said Mikael Thinghuus, ECCO's COO, in a public statement accompanying the opening of a new manufacturing facility in China. "Whereas several competitors phase out and outsource production, we at ECCO maintain that our core is that our own plants are responsible for the main part of the production. The company wants to be the best, not the biggest, shoe producer."[4] ECCO believes that because it maintains extremely tight control of its design, technology, and production processes, competitors require as long as six years to copy an ECCO design and bring it to market.

This of course has a positive effect on ECCO's cash curve. The amount of time the company earns cash for a new design is prolonged, enabling the curve to remain on an upward slope for a much longer duration than the curves for other shoe companies.

Optimal Conditions for Payback from Innovation

Like ECCO, all integrators seek to maintain the greatest possible control over as much of the process as possible. Although they may be willing to use ideas from outside the company, integrators are not willing to wait for or rely on anyone else to deliver those ideas, because they have confidence that they can commercialize and realize their ideas better than any partner, collaborator, or vendor.

It was obvious from our discussion with BMW's Martin Ertl that the company had carefully thought through the risks and paybacks of integration, and that it was thoroughly convinced that integration was the most appropriate innovation business model for BMW and could deliver the greatest return on investment.

Many companies, however, consider integration as the default model and often give little thought to whether it is actually appropriate to their capabilities and the innovation itself. It is important to make an explicit decision on whether and when to be an integrator. BMW integrates in order to stay on top of and drive engine technology advances and to be able to leverage that into the overall vehicle performance. The approach has particular advantages in several other situations:

- When control is necessary

- When the company has world-class capabilities

- When the risks are manageable

- When, simply, the company can

- When knowledge assets have to be protected

- When there is no better (or other) choice

When Control Is Necessary

When the company needs absolute control over product quality, or when the relationships among different components of the product are critical, integration makes sense. Although this would seem to be an issue primarily for high-technology companies, control of quality can be key for a wide variety of other products, including Scotch whisky.

The Balvenie, for example, is a Speyside-based producer of single malt Scotch whisky, and the company controls every aspect of the production of its liquor. This control is essential to ensuring the consistency and quality of the product, and is a distinguishing feature both of the product and of the brand.

What is noteworthy about The Balvenie is that its dedication to centuries-old production techniques is part of its innovativeness. Rather than adopting a variety of modern methods of production, which rely on new technologies to reduce costs or be more efficient, The Balvenie has dedicated itself to traditional methods, making it unique in the industry. In fact, since The Balvenie's first distillation in 1893, the only major change it has made to its production process has been to replace the coal-fired heating element with a steam coil, which took place in the 1960s.

As we learned on a rarely conducted private tour of the distillery, The Balvenie is the only distiller in Speyside that grows its own barley and does its own floor malting on-site. It is the only distillery in the world that also employs coopers to maintain the casks and a copper-smith to look after the stills.

The Balvenie is also one of the few distillers that maintain on-site bottling halls. Most other distillers dilute the whisky and bottle it in facilities located some distance from the distillery itself. When the whisky emerges from the cask—after twelve, twenty-one, even thirty years in the wood—it is generally diluted from "cask strength" of around 60 percent alcohol content to 40–43 percent. When the dilution takes place at an off-site bottling facility, local water is used. Water has a characteristic flavor profile depending on its source, and the unique mix of minerals and other components has a noticeable impact on the flavor of the whisky. By bottling its whisky on-site, The Balvenie is able to use the same water that originally went into the raw spirit and thus avoid changing the flavor and nose it has worked so hard to attain.

The Balvenie was also one of the first producers to recognize that aging in two different types of wood could enhance the taste of a whisky. The Balvenie's DoubleWood was one of the earliest multiwood finished products, a treatment that is now widespread in the industry. DoubleWood, thanks to its early entry into the market, became a major success for the brand.

What does all of this mean for The Balvenie's cash curve? The company's integration approach entails significant start-up costs, and the speed to market is slow (a fifteen-year-old single malt requires, yes, fifteen years to produce). But the integration approach delivers indirect benefits that eventually result in very attractive cash payback. Because Scotch lovers have learned about and value the unique process, the brand has great strength, and the company is able to charge a price premium over other single malt brands of similar age and quality. In addition, The Balvenie employees take pride in the fact that they adhere to traditional distillation methods.

When the Company Has World-Class Capabilities

Integrating is the optimal model for the company that has world-class capabilities all along the innovation process, such as Nokia, the world's number-one maker of cell phones.

As an integrator, Nokia does nearly all its manufacturing in-house, designs its own chips, and manages with unmatched prowess a logistics chain through which flows some 60 billion components each year. The company believes that design and supply chain assembly are its core and distinguishing capabilities.

Nokia's integrated model has also resulted in a robust cost advantage, which places it in good stead when compared to the rest of the industry. In handsets, Nokia has an estimated 20 percent cost advantage over its rivals. CFO Rick Simonson attributes this difference to integration—its in-house manufacturing, sourcing, and use of standard components.[5] Nokia's next generation of cell phones, which will use wideband code division multiple access, or WCDMA, is expected to have a cost advantage of at least 20 percent and as much as 30 percent.[6]

When the Risks Are Manageable

Risk is another factor that potential integrators must consider. Specifically, integration may be the best choice in situations when you don't want to spread risk, or don't need to.

Zara, which is a unit of Spain's Inditex SA, is a maker and retailer of fashionable, reasonably priced apparel. Fashion apparel is a notoriously fickle and risky industry—manufacturers and retailers essentially place big bets, months in advance, about what is likely to be "hot," innovative, and in demand during the next season. As a result, many in the industry do all they can to hedge their bets, hand off the risk to another party, or try to improve their operations and reduce their time to market. Zara, on the other hand, doesn't see the need to avoid the risk at all costs—in fact, the company embraces it. Zara has turned its control over garment factories into a competitive advantage by not only selling clothes but also designing and making them. Amancio Ortega, Zara's founder and the richest man in Spain, says that to be successful, "you need to have five fingers touching the factory and five touching the customer."[7]

Zara is a large operation, with more than 760 stores in 55 countries worldwide, but because it does everything itself, Zara can react swiftly to changing market trends and quickly distribute new products throughout its system. While other companies in the industry take up to nine months to get new product lines into their shops, Zara takes only two to three weeks. Zara shop managers report back to designers on what has and has not sold. This information is used to decide which product lines are kept or altered and whether new lines are to be created. This deep market knowledge enables Zara to get its support activities right and also helps it create new products that can quickly achieve scale.

In its factories, Zara does not focus solely on maximizing utilization to drive manufacturing efficiency. Rather, the company intentionally retains extra capacity to enable it to quickly respond to changes in demand. Rather than seeking economies of scale, it manufactures and distributes products in small batches. Instead of relying on outside partners, it controls all design, warehousing, distribution, and logistics functions in-house.[8]

But there are risks involved in integration Zara-style. The company is highly exposed to the whims of the market and always faces the possibility that something could go wrong in some part of the process, from design through distribution. In Zara's case, however, the risks have been far outweighed by the cash rewards. Zara collects 85 percent of the full ticket price on its retail clothing, while the industry average is 60–70 percent. It also has fewer out-of-stocks of popular items and less obsolete merchandise sitting on store shelves, both of which benefit the brand. And as Zara hones its reputation for stocking styles people actually want to buy, its brand grows stronger, resulting in even more cash payback.

When, Simply, the Company Can

Sometimes a company integrates simply because it can—because it has a unique set of assets or capabilities that can be combined to create new offerings. Nikos Kardassis, head of Merrill Lynch Credit Corporation and head of business development and distribution Global Bank Group, said that this was the case with regard to developing new financial products for the millions of baby boomers in the United States who are expected to retire in the coming years. Merrill Lynch, Kardassis told us, is developing a new Income Management Account product that will combine many elements that most consumers now obtain from several providers and must manage separately—including income-generating instruments, insurance components such as long-term-care coverage and death benefits, equity investment tools, and fixed-income offerings. "To address the needs of people to fund their retirement over twenty to twenty-five years, regardless of what might happen to them, you really need to put a full platform together," he said. "But to do so, you need to have insurance expertise, along with analysis and actuarial skills, as well. But you also must be able to put investment products into that—a mutual fund, for instance, or some fixed-income products. Very few companies are in a position to do all of that, because, for the most part, they're either insurance companies, or they're investment companies. We happen to have all the pieces."[9] In other words, Merrill is doing this as an integrator because it has everything it needs to bring the whole offering together, in a way that few competitors can match.

When Knowledge Assets Must Be Protected

The integrator approach is important to consider in situations when the company does not want to share knowledge or reveal vital secrets. The knowledge in question can be of any type—including all the "ways to protect" we discussed in chapter 3, such as customer insights, trade secrets, and trade dress. If the knowledge is absolutely critical to the success of the innovation, then keeping it in-house through integration is an appropriate move.

Assessing the strengths and merits of knowledge can be an opaque situation, one likely to spark heated arguments. Typically, however, we see managers leap to the assumption that all knowledge must be protected at all costs. They argue that knowledge is critical to competitive success, and they are right. It's just a question of how critical it is and how risky it will be to have it exposed to others.

For example, throughout the 1990s, Motorola was one of the top patent-receiving firms in the United States. This was due to a stated strategy "to get as many patents as we could," according to Jonathan Meyer, Motorola's senior vice president for intellectual property (IP) law. Amid a round of cost-cutting measures in 2000, Motorola's IP strategy shifted its focus to technologies it deemed most critical. While part of this switch was due to cost savings (a global patent for an invention can cost up to $200,000 over its life), a key driver of the strategy shift was an emphasis to focus on the most valuable technologies. As Charles Backof, a vice president in Motorola's technology office, recently said, "Are you ever going to make money off of a battery latch patent?"[10]

When There Is No Better (or Other) Choice

Sometimes a company has no choice but to integrate. This is especially true when it develops, or wants to develop, a truly breakthrough product or service. Sony's experience in developing its Camcorder video recorders is a case in point. "Sony cofounder and chief adviser, the late Masaru Ibuka, was crazy about cameras and 8 mm movies," Teruaki Aoki told us. "But it took one or two weeks to develop the film. So he wondered if we could make electronic-based movies that would play back instantly." The challenge was that some of the basic building blocks required to do this had not yet been created. "To make electronically-based movies, we needed an electronic eye—a sensing device. We also needed a high-density recording mechanism. Then, of course, we realized that high-density tape wasn't available either. So we developed the metal tape, the recording mechanism, and the sensor by ourselves."[11]

Sony spent almost fifteen years and more than $200 million on development, and in 1982 it launched its Camcorder line. By 2001, more than 150 million units had been sold worldwide, with Sony reigning as global market leader.[12] The lesson it took away from the experience was clear: to do something that has never been done before, it may be necessary to do it all yourself. And while the start-up costs can be daunting, getting it right can mean a handsome payback, especially when the payback is not shared with others.

Seagate: A Large-Scale Integrator

Integration may be the classic model, but that does not mean it is archaic, or waning in popularity, or being shunned by the most innovative companies—in fact, some of the most successful innovators on the business landscape, such as Seagate Technology, are integrators. The company has leveraged the approach to race up the rankings of the world's most innovative companies in just over a decade. Seagate has distinguished itself from competitors with its commitment to R&D and integrated product development in an environment where orchestrating might be extremely tempting.

Based in Scotts Valley, California, Seagate is the world's largest manufacturer of computer disk drives, including the all-important "head," the element that reads the information from the disk. Founded in 1979, Seagate shipped the industry's first 5.25-inch disk one year later, a technical milestone that helped fuel the PC revolution.

The disk drive industry is a tough environment, and companies need to work hard to keep pushing the envelope of technical performance. Memory storage capacity, in particular, has increased over the last ten years at a rate that outstrips even Moore's Law in semiconductors.

At the same time, commoditization is always just a heartbeat away. Competitors quickly match capacity increases, driving prices down fast. As a result, many companies in the industry, such as Hitachi Global Storage Technologies and Fujitsu Limited, have scaled back on R&D and other operations, focusing instead on efficiency and speed as the way to become as competitive as possible on price, and have moved toward the orchestration innovation business model.[13]

Seagate's chairman and former CEO, Steve Luczo, spoke with us about the risks and rewards of integration, and how model choice aligns with product strategy. "All manufacturers—and Seagate is primarily a manufacturer—need to be able to control their product strategy," he said. "For us, that means controlling the technology. Because how our head, disks, electronic circuitry, and other components evolve—and evolve together—determines how our product strategy evolves. The manufacturing technology and know-how are intricately tied in to the design. The knowledge of how you will manufacture a head, and of the type of equipment you're going to use to make it, is absolutely integral to how you design the head in the first place."[14]

Luczo does not consider orchestration an appropriate strategy for Seagate. "Let's suppose we decided to get rid of our head business entirely and source from an external provider," he said. "It just wouldn't work. The head and disk interface are becoming increasingly critical to this business. In order for us to have confidence that those heads were being developed along the right technology path, we'd need to have extremely tight integration between our head- and drive-design teams and the manufacturer. But no head manufacturer is going to give us that kind of access to its technology unless we're willing to commit to a long-term deal. And a long-term deal is going to take the form of a long-term purchase contract, meaning we're giving up any cost advantage. So why would we do it?"

Integration also provides Seagate with control over timing, quality, and cost. "We can ramp up production quickly, on our own schedule, so we know the components will be available when we need them," said Luczo. "We can also make sure the quality is good and that the cost stays down, which is critical in a low-cost industry. We'd always felt that, as product cycles got shorter and customer demands for storage increased, this degree of control would be increasingly important. And, in fact, it has become so."

An additional advantage, Luczo says, is the ability to better utilize R&D spending. "Early on, we asked ourselves this question: For every dollar we spend on R&D, how can we leverage that into as much revenue potential as possible? The answer, we determined, was by distributing it across a broad product line, one we designed and fully controlled. We worked to develop heads, disks, and motors that could be used in a range of drives—one-inch, high-end, mobile, and others—rather than in simply one product. This entailed making some huge investments. But because we knew that we had ourselves as a customer, we could make these investments confidently with a three- to five-year time frame, knowing that the payoff would be there and extendable throughout our product line. We couldn't have achieved a comparable result by outsourcing the design and manufacture to a third party. None would have made the commitment of capital or R&D necessary to design technology that could be utilized so broadly. Doing it alone gave us, and continues to give us, a huge advantage."

In Seagate's view, the steep initial dip in the cash curve that comes during development is more than offset by the cash payback generated once the products are launched. And the fact that the investment can be applied to later generations of innovations both increases the cash generated for the initial innovation (prolongs the duration of being in-market and cash-generating) and decreases the investment required for the subsequent product lines.

Luczo also talked about the organizational payback that occurred from being a successful integrator. "It was hard. When I took over, Seagate had some pretty significant cultural issues that we had to address. It wasn't, for most people, a really fun place to work. People really worked hard, and Seagate was typically the leader. But it was tough, because there wasn't any obvious payoff in terms of consistent earnings. It's now a great environment—a winning environment. There's a lot of cooperation, and there's no competition among the major functional organizations, which there was before," said Luczo. "You can't have your own teammates competing against each other, because it's just going to lead to bad behavior for the company. It may optimize a particular functional organization, or a particular subset of an effort, but ultimately it's going to work against the whole. It took people a while to grasp the idea that they really need to think about the company as a whole, rather than about their individual group."

Requirements for Success in Integration

Seagate makes integration look relatively easy, when, in fact, it entails a high degree of difficulty and risk, and requires several capabilities to achieve success, including:

- Ability to change

- Coordination of multiple activities

- Management of relationships with others

Ability to Change

Integration can improve market response time, but it can also have the opposite effect—it can slow a company down. Integrators that get into structural binds may become unable to react to changes in the competitive environment. One reason is that big investments, like those required in integrating, tend to spur actions to protect the investment—both in psychological and operational terms. Management may be unwilling to walk away from or threaten an existing product line that the company has spent huge amounts of resources building. As a result, new opportunities may be ignored or downplayed.

This can happen even at innovators like Intel, where, until recently, any proposed idea that did not support the core PC microprocessor business was pushed aside. Former chairman Craig Barrett used to liken it to a creosote bush, which kills off all the other vegetation around it.[15]

Coordination of Multiple Activities

Integrating also presents a huge coordination challenge, very different from the coordination required to orchestrate. Integrating has its own kind of complexity, which can be overlooked because companies feel so familiar with the approach, but it can be incredibly challenging to manage.

For Zara, the fashion retailer, integration involves managing many activities in order to generate new clothing designs, manufacture them, get them to stores and on shelves quickly, and sell as many of them as it can at the highest possible price.

Zara's supply chain is organized to transfer data quickly and easily among retail, design, manufacturing, and distribution departments. Zara shop managers regularly report to headquarters about what is selling and what is sitting on the shelves. Designers act immediately on the information, determining which product lines to keep or alter and which new lines to create. The designs are transmitted electronically to a network of Zara factories, which use sophisticated just-in-time systems that enable them to put items into production quickly and produce just the required quantities. All new items are priced and tagged before delivery to the stores so, when they arrive, they can be delivered directly to the floor. In an industry that traditionally allows retailers to change a maximum of 20 percent of their orders once the season has started, Zara lets its retail managers adjust an amazing 40–50 percent of their orders. Coordinating all these activities can be a daunting management challenge, especially in an industry that moves as quickly as fashion does.

Managing Relationships with Others

Relationships within the ecosystem may be an issue for integrators. The integrator does not always tap into its suppliers' full capabilities, especially when it deals with them primarily on a cost basis. Suppliers may deliver only to specification and not contribute to innovation in a meaningful way. In the automotive industry, for example, the companies that nurture their suppliers and offer incentives to promote innovation, such as Toyota, have tended to be the most successful companies in the industry. Indeed, many suppliers say that they prefer to work with an orchestrator rather than an integrator. True collaboration usually is more attractive than a cold contract alone.

Polaroid: The Rise and Fall of a Fabled Integrator

A danger for the integrator is that once it becomes successful, it can also become overconfident and overestimate its capabilities—which is what happened with Polaroid, the imaging pioneer.

While many factors were involved in the company's downfall, the problem certainly wasn't that Polaroid lacked the ideas, resources, or opportunities to succeed in the new digital world. Nor was Polaroid caught unaware of the shift from traditional film technology to digital—it had started investing in the area at least a decade before it ran into trouble. And Polaroid wasn't simply overtaken by smaller, faster-moving, more innovative upstarts. It lost out to Sony, Kodak, Olympus, and other large, established companies. The real problem was that the integrator approach that had served Polaroid so well for so long was poorly suited to the new world and reality it faced.

For years, Polaroid had enjoyed what amounted to a monopoly position in instant film and cameras. The company became accustomed to high start-up costs, achieving scale quickly, charging high prices, and supporting products for the long term. It was clear, however, that the industry structure for digital cameras and imaging would be fundamentally different. Competition would be intense, coming not just from traditional photographic companies but also from consumer electronics and computer manufacturers. In addition, Polaroid's engineering staff was conditioned to long product development cycles, protected by twenty-year patents, and its manufacturing processes were vertically integrated, from plastic pellets to finished products.

The nature of the inventions that Polaroid would have to develop also presented differences in the knowledge required. Most of Polaroid's research capabilities were in product-applicable areas such as optics, human perception, and film technology—not electronic digital signal processing, semiconductors, software, and storage technologies. Polaroid had to invest significantly to develop a strong position, and had to keep investing.

And, of course, the risks were high. Since the market for digital cameras was nascent, there was no way to anticipate exactly how and how quickly customers would adopt and react, nor exactly how the company would make money with no film to sell.

Despite all these differences from its traditional instant camera and film business, Polaroid chose to participate in the digital space as it always had—as an integrator, incorporating the fruits of its research efforts into high-quality and high-price "new to the world" products that it made itself. One of the earliest commercialization attempts of its digital technology, Helios, a digital laser imaging system intended to replace X-rays, consumed nearly a billion dollars in investment, including the construction of a 250,000-square-foot manufacturing facility in New Bedford, Massachusetts.

Polaroid's next major attempt to commercialize its digital inventions was the PDC-2000 camera, a product initially aimed at the commercial market and also intended as a platform that could someday enable Polaroid to compete in the consumer market—even though, at the time, Polaroid did not expect that the consumer demand would ever grow very large. The PDC-2000, while technically strong, was late to market and priced too high. Introduced in 1996, it sold at prices from $2,995 to $4,995 at a time when consumer products from other manufacturers were selling at prices below $1,000. One industry analyst commented about Polaroid, "In the future, they're going to have to be a lot faster."[16] Both time to market and time to scale were critical.

Polaroid's first major thrust into the general consumer digital camera market didn't come until late in 1998, five years after competitors' products had begun shipping. Discontinuing the PDC-2000, which had proved to be unsuitable as a cost-effective platform for consumer offerings, Polaroid outsourced its consumer business to Chinese manufacturers, adding minor cosmetic features and selling under the Polaroid brand name. Capitalizing on its trade relations, Polaroid was the first major manufacturer to sell large volumes through Wal-Mart, and it eventually took the number-three position in the market.

Even so, with little value-added and a brand that by that time had been overtaken by a number of others, it simply could not command prices that generated sufficient payback. Caught between declining instant film sales, an inability to generate returns from its digital imaging business, significant continued technology investments, and a weak balance sheet, Polaroid could no longer compete. Choosing to do what it had always done, in a situation where different approaches were called for, resulted in its innovations being too slow, too expensive, and too much for the company to effectively pursue. What had been a promising set of technological innovations, in fact a leadership position, had been lost. Although the company had developed substantial knowledge, its poor choice of innovation business model meant that most of that knowledge could never be converted into cash.[17]

The Polaroid story ended sadly—in bankruptcy, liquidation, and sale of the brand to a licensing company that has slapped the name on a wide variety of products, none of them innovative. The lesson is that, in the right situation, integration can enable a company to become a hugely successful innovator. But integration also has its dangers. Integrators can get cut off from the outside world and become more susceptible to being blindsided by a new technology or competitor from outside its traditional cohort. And by relying almost exclusively on internal talent, integrators often don't take full advantage of skills and ideas beyond their walls. What's more, the resources that once were considered advantages—people, processes, and plants—can become liabilities.

The mightier the integrator becomes, the more diligent it must be about avoiding insularity and inflexibility.

The Necessary Role of Integrators

Integrators play a powerful and important role in business and the global economy. Many of the biggest product breakthroughs have come about thanks to companies that were willing and able to handle the significant risk and huge investment required to integrate. And it is likely that integrators will find solutions to today's most pressing concerns.

Consider what Toyota, Honda, and other automakers have been doing with hybrid and alternative power sources. Both technologies are still years away from maturity, yet these companies are investing heavily and beginning to get some traction with them. Toyota and Honda took an early lead, pumping billions into development of the hybrid vehicle technology and, in Toyota's case, as much as $30 million into advertising.[18] Whether or not you believe in hybrid technology, it's clear that the integrators are trying to make it work, just as Seagate continues to push the envelope in hard drives, and many others do in their own arenas.

Undertaking and managing risk is part of innovation. Integrators assume risks that, when overcome, often result in tremendous cash payback and substantial indirect benefits for the company and help create a better world for all.

FIVE

The Orchestrator

You have to place trust in people who don't
work for your company.

—Don Maynard, senior engineering manager, Whirlpool

O N THE BACK of an iPod, near the bottom of the case, a
legend in tiny letters reads, "Designed by Apple in California. Assembled in China." You might think this is a sign of globalization or proof that Apple outsources some of its manufacturing
operations. However, the label, and hundreds of others like it on products created by companies in all kinds of industries, is evidence of a
more subtle change that has occurred in the way many new products
and services are being developed and brought to market: it's known as
orchestration.

If a company chooses to integrate because it believes it can best
manage the entire innovation process itself, a company decides to orchestrate because it has made the opposite assessment—that it can
best innovate in collaboration with one or more partners. Orchestration is not about control through ownership of the entire process; it's

about management of a network of contributors, all or most of which have a stake in the outcome.

The key to successful orchestration is determining which parts of the process to keep in-house and which to entrust to partners. Usually, the company will choose to employ its own strongest capabilities and look to outside companies to supply capabilities that the orchestrator does not have or is weaker in. But this is not always the case. Sometimes a company will decide to develop an in-house capability to handle an aspect of the innovation process that it deems to be too competitively essential to put in the hands of others. And sometimes a company will ask a partner to provide a capability or deliver an asset that the orchestrator already has but does not wish to deploy for this particular product or service.

Orchestration involves more than simply the bringing together of a set of capabilities. It is the assembly and management of a whole range of tangible and intangible elements—design skills, manufacturing capabilities, a workforce, a brand, a distribution system—into a functioning whole.

The decision about which aspects to keep and which to assign to others must be made according to the four S factors of the cash curve: how best to manage the amount and risk of the start-up costs, how to get to market at the best moment, how to achieve scale as quickly and efficiently as possible, and how to manage support costs to achieve greatest payback.

Orchestration Is Not Outsourcing

The orchestrator and its collaborators may work together in many different ways. While the orchestrator may not necessarily be the one that generates the original idea for the invention, it is the primary "owner" of the idea and the main driver of the innovation process. Although one company generally runs the show, it is possible for two or more companies to be co-orchestrators, with shared responsibility for managing commercialization and realization—the Sony Ericsson partnership in mobile phones is a good example.

The success of the orchestrator rests on its ability to work with and leverage the capabilities of other companies. These relationships are different from the traditional master-servant relationships that the integrator often develops with its suppliers. The orchestrator works in much closer collaboration, inviting its partners to get involved in sensitive and mission-critical activities, such as joint research, joint product design, or joint entry into new markets. This means that all the orchestrating partners must invest time and resources in order to develop and manage the relationship, especially when there are differences in values or cultural norms.

Orchestrating is also very different from outsourcing. The outsourcing company hands off specific activities or processes to others, usually to cut costs or eliminate uncompetitive assets, and does not expect its vendors to participate in the actual process of innovation. Orchestrating involves far more than the purchase of materials or components of a new product—it involves partnering on significant and critical aspects of the process.

With all the attention paid in recent years to the concept of networks and open innovation, most companies have taken steps to get closer to key parties in their ecosystems, especially suppliers. However, many such attempts are superficial at best. While most companies today are fairly skilled in negotiations and vendor management, many are still much more comfortable putting pressure on suppliers than they are considering what is best for both or all organizations involved. Managers may spend a little more time on the phone with their buy/sell partners and speak the language of collaboration a little more fluently, but when push comes to shove, it usually gets rough, and the true nature of the relationship becomes apparent. Although it's easy to spin outsourcing as orchestrating, it takes a lot more than lip service to make it work for real.

Despite the many "faux orchestrators," an increasing number of companies in almost every industry truly understand orchestration and achieve success with the model. The rise in successful orchestrators has been driven by advances in information and communications technologies that facilitate collaboration and by the general globalization of commerce. Another contributing factor is the many highly skilled and extremely specialized companies available to collaborate with, all along the value chain and in virtually every business field. It no longer makes sense to do something yourself as an integrator if collaborators are available who possess skills and capabilities that far surpass your own, who are easily accessible, and who know how to work in collaborative relationships.

Flextronics and Li & Fung, for example, are both companies that were born to help others orchestrate.

Flextronics is a global contract manufacturer that offers turnkey manufacturing services to some of the world's leading electronics companies, and has been Microsoft's partner in the creation of the Xbox. Its services include design, test services, component solutions, manufacturing and assembly, and various other value-added services. The company partners with clients in any phase of the innovation process, from idea generation to realization.

Li & Fung acts as connective tissue between global brands and their far-flung manufacturers. It offers many services, including assistance in designing products, materials sourcing, and logistics management. In effect, Li & Fung provides its clients with a virtual company. It's no exaggeration to say that if you wanted to create a new product and had the capital but absolutely no existing capabilities to produce it, you could contract with Flextronics or Li & Fung, and they could orchestrate it for you. They might even locate the capital for you. All you would have to do is have the idea and sell the result.

The Tablet PC: An Unusual Form of Orchestration

Microsoft chose a form of orchestration for the development of the Tablet PC, a computer that users can control through a stylus and a touch-sensitive screen. "The idea sprang from Bill Gates's vision," said

Bill Mitchell, corporate vice president for Microsoft's Windows Mobile Platform Division. "He has invested heavily in pen-based computing for about fifteen years and has a long-term belief that the lack of truly natural user input is one of the barriers that prevent PCs from being useful to everyone. Humans have been using pen and stylus implements for thousands of years, so it's a logical choice."[1]

The creation of the Tablet PC would involve advances in both software and hardware. Because Microsoft is primarily a software company, it did not have the in-house capabilities to manufacture the Tablet PC hardware itself, nor did it want to develop those capabilities and risk being seen as a competitor by its largest customers. But Microsoft could not simply "throw" the operating system software onto the market—as it essentially can do with Windows—because no hardware platform existed, and Microsoft could not be sure that any manufacturer would make the investment to create one. Moreover, Microsoft needed to design features that were supported by the hardware platforms.

So Microsoft chose to orchestrate and partnered with a dozen hardware companies—including IBM, HP, Toshiba, Acer, and Fujitsu. Microsoft licensed the operating system, but also worked closely with each company to adapt the software to their needs and help them create products—from souped-up laptops to electronic slates—that were differentiated from those of its competitors even though they all operated with the same Microsoft-supplied software. Microsoft was also able to leverage its partners' extensive knowledge of pen-based computing for the benefit of all.

"Microsoft can design highly sophisticated and innovative software," Mitchell said, "but the software is only as good as the hardware it runs on. We have to find a good hardware receptacle and partner effectively with the companies that make that hardware so we can get it to the right customer segment with the right message."

Introduced in 2002, the Tablet PC had a slow start, but sales have grown steadily, and the latest models have won increasingly strong reviews, helping push it beyond niche applications with health-care workers and insurance adjusters. The IBM ThinkPad X41 Tablet, for instance, was an immediate hit with businesspeople.

Mitchell talked about the issues involved in orchestrating the Tablet PC: "One of the most useful lessons is to embrace the differences we have with our partners. Instead of looking at the relationship as a challenge, we embrace the fact that we have the opportunity to come out with many variations of the product. When we have discussions with IBM, for instance, they've got their own experiences, takeaways, and market data. Compaq and HP did the Compaq Concerto, and they also have their own takeaways from pen-based computing. So you're coming at the relationship with different learnings. This can either give you an excuse to walk away because you just can't agree, or you can take the best of the experiences and try to put together a machine that addresses everything."

Working with partners in this way gives Microsoft multiple routes to achieve cash payback. It also provides the indirect benefit of strengthening the company's ecosystem, by giving its partners access to the operating system and enabling them to develop the products they most believe in.

However, there is some tension between those goals, and it took some time for Microsoft to learn how to live without total control over its partners and support them in creating their own distinctive products. In particular, Microsoft wants to ensure that all the pen-based products created by different partners have excellent ease-of-use characteristics and some standardization of features, because these directly affect customer acceptance and thus time to scale and cash payback. At the same time, Microsoft needs to allow its partners to differentiate themselves from one another and create products that incorporate unique features, even if they need special support from Microsoft.

"That's the biggest risk in taking one company's contribution and software and mating it with another company's hardware," said Mitchell. "One company may develop a thumbprint reader that's not supported in the operating system. Another might want to incorporate a built-in 3G radio, but it doesn't get supported in the operating system in the way that the company that's building it expects. There are a lot of opportunities for mistakes if the companies aren't synchronized really early."

Without this synchronization and management of risk, Microsoft's cash curve would never rise above the cash breakeven line because of the lack of required hardware and would only sink deeper as the company poured more into endless product development and support activities. "As we work with partners to build ultra-portables and other machines that are more and more personalized, we realize that we need to do a lot of side-by-side work," said Mitchell. "For instance, with the new generation of touch screens, tweaking the drivers to make the touch screen work really well. If we didn't partner with the OEMs in doing that, the machines and the customer experience would be a lot less satisfactory."

Conditions When Orchestration Warrants Consideration

There is no standard set of criteria that can definitively determine which companies, industries, or inventions are best suited for orchestration. The choice is always situation specific—meaning that what is right for one company, or one new product or service, isn't necessarily right for another. In certain circumstances and situations, however, orchestration should be carefully considered:

- When a capability is missing
- When entering unfamiliar territory
- When you don't want to invest in building capabilities
- When you trust others
- When you want to share risk

When a Capability Is Missing

Orchestration makes particular sense when a company is missing some critical capabilities or assets and when it would take too much time or be too expensive to acquire the missing elements. However, another company must not only have the capabilities, it must also be able to leverage them.

When Entering Unfamiliar Territory

Orchestration can be the best model to use when launching a product in an unfamiliar category, targeting a different customer set, seeking to enter a new geographic market, or expanding a product line beyond its current brand boundaries.

Bath & Body Works (BBW), the $2 billion personal care products and fragrances company that is a staple in malls across the United States, decided to enter new territory in 2003. The company had built its considerable success (it has been called the most successful specialty retailer in the world) by creating personal care products that were based on natural herbal, fruit, and vegetable ingredients characteristic of "heartland" America. The reasonably priced products were sold in BBW shops whose decor echoed the heartland theme and that appealed to middle-market female consumers who were looking for pungent fragrances and good values. The company grew from fifteen stores in 1990 to more than thirteen hundred in 2000. And from 1993 to 2000, BBW posted a remarkable compound annual growth rate (CAGR) of 51 percent.

But in 2000, BBW's sales began to slow, falling 19 percent from 2000 to 2002, and it became clear that something had to be done. A new CEO, Neil Fiske, was brought in. He decided that the brand needed to be revitalized and the way to do so was to broaden its offerings to include more innovative and premium products and brands. To do so, he chose to orchestrate—partnering with a variety of creators of personal care products that could offer existing customers more choice and attract new buyers as well. BBW, which had been dominant in cosmetics, was now expanding into the more complicated and risky territory of "beauty."

"Great brands are about things that you believe in with great intensity," Fiske told us. "We can't replicate that on the inside of the company, but we can partner with people who have it."[2] So BBW pursued partnerships with a number of suppliers of specialty goods—including C. O. Bigelow, "The Oldest Apothecary in America"; L'Occitane, the French retailer; Goldie, a small producer of color cosmetics; and American Girl, the company that offers educational dolls, books, and accessories—to create innovative new lines designed to be sold exclusively at BBW.

Partnerships with well-known brands and products help the cash curve by cutting the start-up costs involved in development and also by reducing the support costs that would be needed to build and support an unknown brand. After Fiske began moving BBW toward orchestration, in the first quarter of 2003, BBW achieved earnings growth in eight of the following nine quarters.

When You Don't Want to Invest in Building Capabilities

By orchestrating, a company may be able to avoid making a large investment in creating capabilities needed in some part of the process. This may be because investment can have a more positive effect on payback if it is applied elsewhere. For example, a company might choose to have partners design, build, and manage a specialized production line to reduce prelaunch investment, but invest heavily in marketing and promotion in order to achieve scale as quickly as possible. Avoiding investment in one area can also be particularly important when capabilities are missing in other parts of the process but are not available from another company or source, and there is no choice but to build them internally.

When You Trust Others

Automakers, which have traditionally been integrators, now orchestrate for the creation of some models. Cars are complex products, composed of tens of thousands of parts and components, assembled in a process of many steps. Development of a new model involves significant investment, in both the start-up and the support phases, and there is such intense competition that there can be little certainty that the new product will reach the scale required to achieve payback.

In fact, because the industry has become so complex and the number of capable supply chain partners so numerous, automakers have increasingly turned to orchestration to help them reduce their start-up risk and increase speed to market.

This is why BMW, a company that is a die-hard integrator when it comes to engines, chooses to orchestrate for the creation of other important elements, including its "active steering" system. As Martin Ertl at BMW explained to us, "Our definition of innovation is that it is a valuable novelty for the customer with a market success for the company. So, we must think about what we have to invest, and what we can get in sales, for each innovation. Once the active steering system was budgeted, we decided it would be best to work with external partners so we could focus internally on our strategic core areas. The engineers then have to come up with a plan for how to partner, how to set the gates to control the synchronization points, how to meet the budget, what kind of capacity they have in-house, and what capacity they need from outside. The advantage when you collaborate with a supplier is that you make them stakeholders in the overall process. It is their baby as well as yours."[3]

When You Want to Share Risk

Orchestration works well when companies want to share risk. Boeing, for example, chose to orchestrate and to share the start-up costs with its suppliers in the development of its 787 aircraft.

In December 2003, Boeing announced that it would produce a new aircraft, the 787. Boeing had considered many designs, including a reconfigured 747 and a high-speed sonic cruiser, but had finally decided to move forward with the 787, an aircraft project that would focus on efficiency rather than on large passenger capacity or exceptionally high speed. (The plane was originally dubbed the 7E7, with the *E* standing for efficiency.) The 787 would be efficient to operate—in terms of both fuel consumption and maintenance costs—and efficient to manufacture, as well.[4] As Walt Gillette, Boeing's chief engineer for the 787, put it, "We need to take commercial airplanes into the mainstream of manufacturing."[5]

However, market analysts and potential customers had their doubts about the plane. Continental Airlines chairman and CEO Gordon Bethune dismissed the 787 as a program cooked up by "a bunch of accountants."[6] Byron Callan, a specialist in aerospace at Merrill Lynch, said, "If they whiff on this one, I really do think they'll have credibility issues with their suppliers."[7]

For its earlier models, Boeing had contracted with suppliers to deliver various elements of the aircraft according to Boeing's detailed specifications and complete blueprints. For the 787, however, Boeing decided to change the relationship with its suppliers, working with a smaller number of partners and providing them only with general parameters for a variety of large components, including the entire wing, and relying on them to propose designs and then produce and prepare the parts.[8] Because there are fewer components, Boeing is able to do final assembly more quickly—in as little as three days, in comparison to a month for the preceding model, the 767—which means that it will tie up human resources and factory space for a shorter period of time.[9]

This arrangement requires that the suppliers take on a significant percentage of the cost of development and participate far more substantially in the development of the wing design. Boeing's interim CEO, James Bell, announced in April 2005 that of the $2.3 billion Boeing would spend over twelve months, $600 million would be contributed by its collaborating partners, suppliers, and others involved in the project.[10]

This does not mean that Boeing has reduced its start-up costs by $600 million, because there will be added costs associated with implementing the orchestration model—particularly in developing a design that can accommodate the assembly of the various subcontracted modules. However, Boeing's prelaunch cash investment is substantially reduced, which means that Boeing will be able to achieve payback more quickly than if it had shouldered the entire risk itself.

The orchestration of the 787 has also brought Boeing indirect benefit, particularly by strengthening its ecosystem. In fact, Boeing's major suppliers are working hard to ease friction between Boeing, Airbus, and their respective home governments. As Marshall Larsen, chairman and chief executive of Goodrich (which supplies evacuation slides and landing gear for both companies), said, "The last thing the industry needs is a trade war."[11]

The Payback and Indirect Benefits of Orchestration

Because orchestrators share some of the start-up costs with their partners, and also share some of the profits, it might seem unlikely that the orchestration model could ever deliver the payback that integration does. However, the size of the return for companies following either model is largely dependent on how skillfully and successfully the chosen model is executed.

The main reason to choose orchestration is its impact on the three types of risk involved—operational, technical, and market. Orchestration can reduce the operational risk by enabling the orchestrator to take advantage of the superior operating capabilities of its partners. It can minimize the technical risk by tapping into technical skills, secrets, and knowledge it does not possess or cannot develop. And the orchestrator can lessen the market risk by working with partners that have experience and success with particular customer segments, distribution channels, or geographic markets.

The cash implications of orchestration go well beyond reduction of the up-front investment. Orchestration also creates more flexibility to make changes in the product during commercialization and realization— in response to shifts in the market or to moves made by competitors— because the orchestrator is not locked into the capital commitments it might have made as an integrator.

The orchestrator also creates more freedom for itself to deploy resources of talent and capability to other projects that may become important. In fact, apart from one or two key functions that are kept in-house, orchestrators may come to think of their capacity as almost infinitely expandable. Managers become so expert at dividing up responsibilities, assembling networks, and coordinating activities that the cycle from idea generation to realization can be drastically shortened.

Acquiring Knowledge

Orchestrating always involves knowledge sharing as well as resource sharing. By working with partners to create the Xbox, for example, Microsoft is gaining valuable knowledge about the "digital home" of the future, the role its technologies might play there, how consumers interact with advanced electronics, and how to design and develop hardware platforms.

Knowledge sharing also has its risks. First, if the orchestrator comes to rely completely on a partner for certain aspects of its operation, the orchestrator may reduce its own in-house knowledge of that aspect. This may not be a problem until the day the orchestrator decides that it wants to bring that aspect back in-house (or, for some reason, is forced to) and discovers that it doesn't have sufficient knowledge to do so.

Second, the orchestrator may, in the course of a collaboration, share so much knowledge with its partners that it creates competitors. The partner may pass on the knowledge (potentially, but not always, legally and inadvertently) to the orchestrator's competitors in the course of partnering with them. Or the partner itself may set up shop in the same market space. In China, New Balance, a leading footwear company, has had to deal with both competition from a former partner and apparent intellectual property leakage to a new competitor.[12]

Enhancing the Brand

If the orchestrator has a strong brand, orchestrating may enable the company to leverage it in new categories or markets, but orchestration can also have a negative effect on the brand. This is not necessarily because consumers know or care whether a company does everything itself or works with partners to create its new products and services. Orchestration can only harm the brand if some aspect of the process is poorly handled by one or more partners in a way that is visible to customers.

If, for example, the wing of the Boeing 787 had problems in assembly or in maintenance, airline customers would certainly notice. They would know that the wing had been developed by partners in collaboration with Boeing, and would likely blame Boeing for poor management of the orchestration effort. This could easily shake the customers' trust in the brand and make them think twice about placing orders for more 787s, which would have an obvious impact on payback for Boeing. However, if the collaboration results in a wing assembly that makes maintenance easier and faster, and thus speeds servicing of the planes and also minimizes downtime, Boeing will be applauded for its ability to orchestrate and the brand will be enhanced.

Strengthening the Ecosystem

The orchestration model requires that companies work more closely, and in a different way, with people and organizations in their ecosystem than does the integrator. When orchestration is well managed, the relationships among partners become stronger and more productive in many ways. Knowledge sharing becomes a way of life. The coming together of different skills and capabilities, attitudes and behaviors, generates excitement, new ideas, and new opportunities. As Bill Mitchell of Microsoft put it, "You union with your partners, take the best of the learnings, and put together a product that addresses everything you've learned from these different perspectives."[13]

Orchestrating may also threaten your ecosystem. The orchestrator's established suppliers may not have the interest or the ability to function successfully in a collaborative partnership, and may be upset if they are not invited to join in the effort, or may see a decline in business because of it. As a result, the orchestrator may lose favorable terms with that supplier or even watch as the supplier allies itself with the orchestrator's competitors. The orchestrator may also lose touch with certain constituencies—such as a distribution network—because one of its partners is responsible for managing that group. In the worst-case scenario, a partner achieves greater payback and more indirect benefit from the collaboration than does the orchestrator itself. The partner becomes the dominant member of the alliance and may have sufficient power to take over the product or service itself, cast the orchestrator out, or even put it out of business.

Energizing the Organization

Orchestrating allows companies to pursue ideas they previously couldn't, helping fuel and foster creative energy. In addition, it can force organizations to break out of traditional patterns of behavior—reexamining their people and processes, and preventing dynasties from forming. For both these reasons, as well as others, orchestrating helps create more efficient and flexible organizations. The orchestrator is more nimble and finds itself able to pursue more possibilities than before. Once a company has successfully built the capability to orchestrate, it can do so again with greater speed and facility—just as Microsoft was better at orchestrating the Xbox 360 after it had learned the ropes with the original Xbox.

There is an organizational downside, as well. Organizations may begin to lose key people and capabilities. Imagine walking through an office building and coming across the "ghost town" where the design (or logistics or branding) group once sat. It can be a powerful visual reminder that orchestrating can sometimes result in a weakened organization.

Whirlpool Discovers the Upside (and Downside) of Orchestration

Orchestrating can be extremely rewarding as well as very challenging, as Whirlpool discovered when it turned to the orchestration model to create the Gladiator GarageWorks storage system.

In 2001, a small team of sales and marketing people at Whirlpool proposed a line of appliances and storage systems for a new market for the company: the garage. It was an intriguing opportunity for Whirlpool, in particular because it targeted both a new "room" and a new demographic for the company: men, rather than women.

An overriding consideration was to develop the project as quickly as possible. While the various products were new and likely to be leaders in the category, it was obvious that it would be important to be first to market. Establishing a strong brand position, securing retail floor space, and gaining valuable product and consumer experience were going to be key to the ultimate success of the products.

In addition, funding was tight—the team was told to commercialize the opportunity with as few start-up costs as possible. Starting off with a $50,000 budget, Whirlpool developed prototypes, ran focus groups (in garages, with free pizza), and traded products for engineering labor.

During the commercialization process, the development team developed a distinctive "look and feel" for the Gladiator line. It wanted a look that was upscale as well as tough, not unlike the combination of qualities associated with a Harley-Davidson, and settled on a cladding material of heavy-duty "tread plate" of the type used on fire trucks. The group also began exploring potential brand names, first considering Quadrant, which suggests organization and technology, but ending up with the more powerful, masculine Gladiator. It would be Whirlpool's first new major brand in almost fifty years.

In September 2001, after receiving very positive reactions from the focus groups and an Internet survey, the Gladiator team received $2 million in funding to push the project ahead. But the funding came with a caveat: if the Gladiator team couldn't create a line of products, attract sufficient customers, and generate significant revenue by the end of 2002, the project would be terminated.

Whirlpool typically operates as an integrator, with in-house design and manufacturing, but that approach would have cost too much and taken too long for the creation of Gladiator. "When we think of making something ourselves," said Todd Starr, head of product development for Gladiator GarageWorks, "we think in terms of thousands of units a day, not thousands of units a month or a year. We considered making some of the products ourselves. After all, we buy a lot of steel. We bend a lot of metal. It seemed like we ought to be able to do it. But in the end, it didn't make sense to do it in-house."[14]

So the Gladiator project team decided to work with partners to create Gladiator. But over the years, Whirlpool had built an ecosystem composed of vendors, not collaborators, and now had to identify and develop relationships with a whole new set of companies. The Gladiator tooling, for example, was sourced from an overseas supplier that could deliver the tools in a third of the time, at a third of the cost, of the suppliers that made tools for Whirlpool's other product lines. The team also chose suppliers that could provide design capability in addition to production capacity, so they would not have to rely on the already stretched in-house designers and engineers at Whirlpool or get involved in bringing new ones on board. "It would have taken a year just to hire the engineers," said Tom Arent, the Whirlpool general manager in charge of the effort.[15]

Whirlpool also needed to innovate in how it went to market with the Gladiator system. The vast majority of Whirlpool's existing products were primarily sold to consumers through retail channels, but Gladiator would have to be different in this important respect.

First, while some volume was going to come via retail sales, the product and concept needed to be properly displayed in order to make the fullest positive impact on the customer. So Whirlpool made a deal with Lowe's—which had a very experienced sales force that could talk about the Gladiator line knowledgeably and demonstrate it to consumers—to be the exclusive launch partner. In exchange, Lowe's offered the display space and configuration that Whirlpool felt would best showcase its offering.

Second, many consumers were going to want the Gladiator products installed for them, which was not generally an issue with other Whirlpool products, like a microwave or a washer. Whirlpool's builder sales force could handle new-construction opportunities. But for installation into existing homes, Whirlpool partnered with California Closets to leverage that company's expertise in installing home storage and organization systems.

Gladiator was tested in selected stores in Charlotte, North Carolina, in the early fall of 2002 and was rolled out to 850 Lowe's outlets across the United States in November, coming to market almost exactly a year after being funded. That's an extremely fast time to market in the appliance industry, where creating a new product usually takes three to five years. The start-up section of the cash curve was shallow, and both the time to market and time to scale segments were short. "I have never seen a product category where we have had a more positive, strong reaction so early in an introduction," David Whitwam, the CEO, said in a conference call with investors.[16]

"When you operate as the orchestrator, as we did here," Todd Starr said, "you're not going to garner the sort of margins that you otherwise would." So although the prelaunch shape of the Gladiator curve looked good, it came at a trade-off in the shape of the postlaunch curve. But Starr added, "I don't know, though, if there would have been a whole lot of interest at our factories in doing something so small."[17] In other words, Gladiator probably would not have gotten off the ground if the team had tried to create it using the integration model.

Internally, the results have been powerful. "Our innovation effort—and Gladiator's success is part of that—is the single most energizing thing we've ever done in Whirlpool," said Whitwam.[18] The experience has served to ignite even more innovation within the company. Whirlpool now has several hundred innovation projects in its pipeline, far more than at any other time in its history. It is reallocating large portions of its capital spending toward innovation, and the learning from Gladiator is helping it get more from the investments. The team also has helped the company develop a new set of skills, ones that they are transferring across their set of products worldwide.

There were also challenges. Starr explained, "When our engineers were looking for suppliers, they didn't have the organizational support they needed. But that has changed. We now have a person in finished goods procurement whose mission in life is finding Gladiator suppliers. But it took some time to get there and it was painful sometimes. We had one vendor who knocked their head against the wall for several months, invested a fair amount of money, but just produced a lot of scrap, had poor run-rates, and finally cried uncle and said, 'No more.' Another company, that was making the hooks for us, wanted to raise prices on us. We actually had to replace almost all the companies in our regular supply base, which was disruptive and a huge distraction for us. It's important to get the supply base right as soon as you come out of the chute. Many of our existing ones could not work with us effectively, given what we were trying to do. That was a huge challenge for us."[19]

Orchestration Is the Riskiest Model

Although orchestration can deliver payback and indirect benefits, as it has for Whirlpool, the orchestrator also takes on or exposes itself to the highest level of risk among any of the three approaches. Working in close synchronization with other companies entails risks and challenges that are different and less familiar to many companies than managing a project completely in-house. After all, since the early twentieth century, the discipline of management has largely been about finding ways to increase control and reduce risk, and the integrated organization has been seen as the best way to accomplish both goals.

So while it's important to recognize the potential benefits of orchestrating, it's necessary to point out that it presents real and serious challenges. Orchestrating requires a fundamentally different set of skills and capabilities than many companies possess, and there is a much wider variety of "things that can go wrong."

Breakdowns in Execution

When Sony Ericsson was formed in 2001, the media hailed it as the "perfect marriage," because it brought together Sony's deep experience in consumer electronics and Ericsson's deep knowledge of wireless technology. The new venture set its sights on surpassing Nokia as the market leader within a few years, a goal more than a few analysts thought was obtainable. Instead, the venture initially saw its combined market share plunge to 6 percent from 12 percent. What happened? Simply put, the orchestration proved to be more difficult than the companies expected and proved impossible to manage sucessfully in a short period of time. Press reports cited delays in releasing new products, quality problems, and a loss of focus on consumer "wants."[20] The venture did not reach profitability until its second year of operation and has only now started to show its full potential. Nokia remains in the number one spot.[21]

There are many reasons why execution is so challenging for orchestrators. To start, it's hard to align incentives and rewards. Companies need to think carefully about sharing the wealth, dividing the costs and investment, and allocating risk. As Ad Huijser, executive vice president and chief technology officer at Philips, explained to us, "The alignment of risk profiles is always difficult. For the Senseo coffee machine, for example, we teamed up with a coffee producer who had already invested in the development and production of the type of coffee-pods that were crucial for this new coffee brewing concept. Without that, it would have been much more difficult to find someone to work with and make the necessary investment when it was needed."[22]

Another execution challenge is that network-enabled collaborative relationships like those required for orchestration can be expensive and difficult to monitor and track. It takes a keen understanding of best-practice models, sophisticated tools and technologies, and excellent communications. Say one of your partners' manufacturing plants is damaged in a fire. It may take you, as an orchestrator, longer to hear about the situation and its implications for the collaborative project than it would if you were an integrator.

Hollowing Out

Once a company has chosen to orchestrate, it's difficult (though not impossible) for it to go back and reintegrate. Schwinn, the iconic American bicycle manufacturer, painfully learned the danger of allowing orchestration to hollow out the core. At its peak, Schwinn owned almost 20 percent of the U.S. market and produced hundreds of thousands of bikes a year in five factories. The Schwinn name stood for cutting-edge innovation and unmatched quality. It was the bicycle that all American children dreamed of owning.

What happened? There were many missteps, but one of the biggest was the company's decision to shed its design and manufacturing activities and focus on marketing and brand management. Schwinn's management at the time felt that the company's legacy of innovation and customer relationships were the most valuable assets, so they began to send more and more production to a small supplier in Taiwan, ironically named Giant Manufacturing Corporation.

Schwinn did not ask for a stake in Giant even as it handed over four-fifths of its bicycle production to its partner.[23] Gradually, the supplier gained knowledge and expertise, began to develop its own new products, was able to cut costs, and eventually became a far better bike maker than its partner. Giant is now one of the world's largest bicycle companies. Schwinn is nothing more than a brand name owned by a diversified consumer-products company.

Schwinn discovered the hard way that orchestration can look very tempting and seductively easy—let your partners do a lot of the hard work while you sit back and reap the rewards. But there is substantial risk lurking beneath the attractive surface. Unused internal capabilities can quickly atrophy. Collaborators that are meant to facilitate the process can get in the way. Partners may unexpectedly become competitors.

When executed properly, however, orchestration can enable a company to take advantage of a far broader range of skills and assets than it could ever develop internally—and deliver successful inventions, with cash paybacks, it might never have been capable of creating on its own.

The Licensor

We profit from the license fee and a guarantee
that we always get the lowest price.

—Karl Weinberger, head of research and
    development, Schindler

A LTHOUGH INTEGRATION AND ORCHESTRATION are
the dominant innovation business models, there is a third
approach that has long been underutilized—even shunned—by many
companies but that is increasing in popularity as people realize the
impact it can have in generating payback: licensing.

A licensor is the primary owner of the spark of the innovation (although it may not be the originator of it) and sometimes of its commercialization, but has limited involvement in the realization. However,
some licensors specify exactly how their intellectual assets are to be
managed in the market so as to ensure certain standards of quality, performance, and consistency of brand (if their brand name is involved).

The licensor uses the business system of its licensee to avoid the cost and effort (and impact on its cash curve) involved in bringing its idea to market and in return gives up a substantial part of the potential revenue (and payback) upside. Some licensors develop close relationships with their licensees, so they can take advantage of new knowledge gained through realization and apply it to further improvements. Other licensors, however, are more laissez-faire and, once the asset is licensed, have little material involvement with it. Many who have jumped on the bandwagon of trying to monetize their existing patent portfolios fall into this category.

The Misunderstood Model

Despite the attractive potential payback to be gained, licensing still conjures up a variety of images to businesspeople, many of them negative.

Some people think of licensors as barely legal profiteers whose companies are staffed by wily patent attorneys and cutthroat litigators, and who would prefer to extort crippling royalty payments from their often unaware and unwitting licensees or attack their rivals with infringement suits rather than engage in the noble battle of the marketplace.

And there have been just enough cases of real or perceived voracious licensing to lend some truth to the largely inaccurate stereotype. Rambus, Inc., for instance, is a licensor that has found itself embroiled in high-profile legal battles in recent years.

In the late 1980s, Rambus founders Mike Farmwald and Mark Horowitz developed ways that chip interface designs could significantly improve the speed of data transfer from a computer's memory to its microprocessor. Rather than try to raise the $1 billion they estimated would be needed to build a chip fabrication plant, they decided to patent and license their designs to the makers of video game consoles, TV set-top boxes, PCs, laptops, network switches, and other such devices.[1]

Since its founding, Rambus has attracted many customers to its intellectual assets but has been involved in many cases of contentious patent litigation. Its critics claim, for example, that Rambus patented one of its chip designs while participating in an industry group that was developing a set of standards that could apply to new products in the industry. Rambus also provoked the ire of industry participants when it staked controversial patent claims to some alternative chip design technologies. A U.S. federal court initially found Rambus guilty of fraud in failing to disclose its patent applications to some alternative chip design technologies, but the decision was overturned on appeal, and Rambus has been able to successfully claim patent infringement against several companies.[2]

Even so, Rambus has seen its brand image and ecosystem relationships suffer. In the short term, this may not matter, because Rambus owns the technical standard and thus generates a healthy payback. In the long run, however, Rambus may suffer a backlash—in the form of an inability to partner successfully with others or the reluctance of customers to do business with the company—that will affect payback.

Other people think that licensing is about selling a brand name and logo. Fashion designer Pierre Cardin helped create this perception. He was regarded as a highly innovative designer, who used vinyl, silver fabric, and outsize zippers in his futuristic designs such as the "bubble dress," and was the first haute couturier to create a line of ready-to-wear clothing. In 1960, he began to license with the intention of extending his innovativeness to other products. But it soon became clear that he was licensing nothing more than his name and that he had no material involvement in the commercialization of the more than nine hundred items that eventually bore the Cardin logo—including frying pans, crockery, aftershave lotion, floor tiles, sardines, restaurants, cars, boats, cigars, hair dryers, wine, luggage, pens, and orthopedic mattresses. "If someone asked me to do toilet paper, I'd do it," Cardin told a reporter for the *Independent*. "Why not?"[3] Soon enough, there was very little left—neither the innovations that first marked Cardin nor significant residual brand value.

Many companies think of licensing as a process of leveraging unused patents, and this can be a highly valuable activity. In 1990, for example, IBM began to license its unused patents, and its royalties jumped from $30 million a year to over $1 billion annually by 1999, reaping about $950 million in 2005 (about 1 percent of total revenue but about 15 percent of net income, for 2005).[4] Although we encourage companies to pursue this path, our interest in licensing is as a primary means of commercializing a new idea, not for dusting off forgotten patents and recouping investments made long ago.

Dolby: A Brilliantly Innovative Licensor

Ray Dolby, founder of Dolby Laboratories, Inc., is known as a great engineer and inventor, but he also is an incredibly canny business strategist and a model licensor. Most people will know his name or the company's technologies, which are used to improve audio quality for recorded music and movies. Indeed, Dolby Laboratories, founded in 1965, has been able to establish prominent leading positions in music recording, audio cassette players, movies and theaters, home theater and DVD, and, more recently, high-definition and digital television. What's lesser known is that Dolby Laboratories' success is largely the result of the strategic decision it made more than thirty years ago to become a licensor.

Dolby Laboratories' original objective was to develop electronic systems that would reduce the background noise of audio recording tape. A relatively new technology, analog audiotape provided much more flexibility than traditional phonograph discs because it could be started and stopped and could record multiple tracks on one tape. It had one major drawback, however: audiotape suffered from a very distinct background hiss that marred the sound of the performance.

In 1965, the first year of the company's operation, Dolby Laboratories introduced Dolby A-type, a new audio compression and expansion technology that dramatically reduced the hiss without any discernible side effects. Ray Dolby believed that the primary market for the A-type would be professional recording studios, and he felt that the only way to achieve the proper quality and performance from his system was to incorporate it into recording hardware that Dolby itself manufactured—as an integrator. Dolby was right. The Dolby A-type recording device soon became the de facto standard for audio professionals.

Dolby was not so sure about the best innovation business model to choose to penetrate the consumer market, however. Dolby knew that he had neither the ability nor the capacity to manufacture enough consumer playback devices for even a portion of the global market—the start-up costs would simply be too steep and it would take too long to achieve anything like ubiquity. Dolby also doubted that enough consumers would care about audio quality and performance to be willing to pay a premium for audio equipment incorporating the Dolby System, which meant that the product might never achieve sufficient scale to get payback.

So Dolby decided to license the noise reduction technology—which is essentially a design and a set of specifications for the electronic circuit that is the heart of the system—and charge a royalty for each unit made with it. By becoming a licensor and partnering with established players rather than competing against them, Dolby was able to greatly improve not only his time to market but also the postlaunch cash curve: he could leverage the manufacturers' market share, penetrate the market faster, reduce his support costs, and maximize his revenue and profits with far less risk.

Not everyone was convinced it was the right move. Potential investors told Dolby that he should forget about licensing and become an integrator for consumer products just as he had for the professional market. He resisted the idea, arguing that the company might grow more slowly if it couldn't secure outside financing, but that the growth would be profitable and consistent.

Dolby was right about that, too. Dolby Laboratories has enjoyed tremendous success by creating, securing, protecting, monitoring, and improving its intellectual property. "I don't go into any area that I can't get a patent on," said Ray Dolby. "Otherwise, you quickly find yourself manufacturing commodities."[5] Dolby Laboratories holds 944 individual issued patents and has over 1,500 pending patent applications in some thirty-five jurisdictions throughout the world and collects royalties from hundreds of licensees, including equipment manufacturers, content producers, studios, and theaters.

Dolby has also continued to act as an integrator for the development and manufacturing of equipment for the professional market. The company has very successfully employed two different innovation business models, each one chosen for its ability to reduce risk and deliver the highest payback.

In early 2005, Dolby Laboratories went public, earning some $495 million for its founder on the first day of trading.

Conditions When Licensing Makes Sense

If the cash payback can be so lucrative, why do so many companies hesitate to commercialize and realize their ideas using the licensor model? Some organizations continue to labor under the general misconceptions we've described or simply can't see how the model might be useful for them in particular.

There are also organizational barriers to licensing. Some companies simply may not be able to accept that one of the assets they have developed in-house might be more successfully commercialized by somebody else. It's their baby and they don't want anybody else messing with it. In addition to the emotional issues, there may be no tangible incentives for the people who are closest to the intellectual asset to try to leverage it through other companies. For instance, there may be no designated organization that has the responsibility for supporting a licensor approach, so the process is just too unknown and arduous for most managers to attempt.

There may even be a kind of penalty that results from a licensor effort. This can happen when the central organization sees the potential of licensing but then requires the business unit that owns the asset to provide the resources to make it happen—giving no budget credit, revenue percentage, or any other reward in return. Also, licensing an idea may generate lower revenues (even if the margins are better) than commercializing and realizing it yourself. For executives who are granted power and receive compensation according to the size of their organization, licensing may look unattractive. Given such circumstances, it is not surprising that many inventions are either pursued more directly via integrating and orchestrating or overlooked altogether.

In many situations, the licensor model makes sense and is likely to deliver payback, including when:

- The company does not have the capabilities to commercialize the innovation and *can't or doesn't want to commit resources* to acquire the capabilities. Robert Bosch, one of the largest and most innovative auto suppliers in the world, licenses an advanced "silicon micromachining" process it pioneered, because it wanted to establish this key technology as an industry standard with a reliable and wide supply of dedicated equipment for high-volume production. While it had already incurred the start-up costs to develop the technology, Bosch realized that it would have to invest considerably in the equipment development for only a small number of machines for its own use. "The basic process took about five years for us to develop, but we decided that it wasn't high-impact enough for us to manufacture the equipment by ourselves, so we turned to licensing," Dr. Siegfried Dais, deputy chairman of the board of management at

Bosch, told us. Consequently, Bosch focused on the production of miniaturized components based on the technology. In order to utilize available equipment design expertise, Bosch opted to license to companies like Surface Technology Systems, which now manufactures deep-trench silicon etching machines that use the "Bosch process."[6]

- The invention can *create critical mass* or lead to the adoption of a beneficial standard. As we discussed in chapter 3, Sony is seeking to set the standard for the next-generation optical disk technology with its Blu-ray disc technology and is battling a Toshiba-sponsored HD-DVD technology to do so. Even though Sony's technology is widely acknowledged to be a breakthrough, it is still not clear that it will prevail, and even less likely that it will completely vanquish Toshiba's lower-cost, lower-tech offering. For both players, licensing is a key element of their strategy to win.

- Competition *can be transformed into a royalty source.* Procter & Gamble (P&G), recognized as one of the world's most innovative companies, takes an unusual approach to managing its innovation assets by licensing its technologies, sometimes as early as three years after the technology has been patented, commercialized in a product, and realized. Even if the innovation has not been commercialized, P&G will not let it sit on the shelf for more than five years before considering it for out-licensing opportunities.[7]

  Although this approach may appear to be counterintuitive, the early availability of a license to a successful P&G innovation actually puts the company's competitors in a thorny situation. If P&G launches a new product that catches on, the competitor will likely want to create a similar offering of its own. If P&G licenses to one competitor, other competitors may be pressured to improve their products because the license has strengthened the product category. A workaround of P&G's patents may be time-consuming and expensive, or may produce an inferior result. A redesign that does not move away from the patent claims will also expose the company to a patent infringement suit—which can take time and resources to defend (and is very risky, since there will be penalties if the claimant loses). Of course, the competitor can attempt to create an alternative approach to provide similar or competitive performance, but this path is uncertain, and P&G will also be pushing ahead at the same time.

Alternatively, knowing that P&G's patents will be available to license, the competitor might get to market faster, and at lower cost, simply by waiting to acquire a license. Meanwhile, as the competitors try to figure out their next move, P&G is preparing a slew of new products that will supersede the current invention anyway. As Jeff Weedman, Vice President of External Business Development at P&G, told us, "My goal is to license our innovations to our competitors for one dollar less, and one day sooner, than they can do it themselves."[8]

P&G's actions are designed to improve the cash curves of its products. By enabling competitors to embed its intellectual assets in their products, P&G gains a "double" cash payback on products it has already realized. Licensing may also reduce the motivation for competitors to invest in basic research that would lead, as opposed to follow, the licensor's technology.

Not only does this strategy put P&G's competitors on edge, it also keeps the company's own people from resting on their laurels. They have to continue to innovate, since their inventions will soon be made available to the world. This awareness tends to shorten the time-to-launch section of the cash curve, which often reduces the required start-up costs as well. It may also motivate P&G people to step up the level of innovation because they are, in effect, facing the prospect of competing with themselves.

The Payback and Indirect Benefits of Licensing

Many companies shy away from licensing not because they misunderstand its nature but because they assume the cash payback from licensing an intellectual asset will be too small to warrant the effort and resources involved. But licensing generally involves a smaller prelaunch investment than other models, and licenses typically generate 75 percent to 95 percent margins, which means that a $15 million to $20 million licensing arrangement can have more value than a $100 million revenue business, with far less risk and far less financial and human capital investment. And the cash often starts flowing more quickly than if the innovating company had tried to commercialize and realize the new product by itself.

Knowledge Acquisition

As P&G has shown, licensing can be used as a strategic weapon, not just as a way to generate easy money through collection of fees.

Schindler, the Swiss elevator manufacturer, brilliantly uses licensing as a way to help fund the acquisition of new product-applicable knowledge that can generate payback in three ways: through license fees, through application to Schindler's own elevator products, and through eventual application to other, nonelevator systems.

Schindler developed a new elevator control technology called Modular Shaft Information System (MoSIS) that greatly improves the ability to monitor and control the location of high-speed elevators with large travel heights. The system is so precise that it can bring an elevator to a stop to within half a millimeter of a desired location. Not only does MoSIS improve operation and safety, it improves elevator availability. The system is able to monitor and compensate for many variables, including stretching and tautening of the elevator's cables, as well as building shrink—a physical settling of the building that can slightly shift the location of each floor.

The heart of MoSIS is magnetic measuring technology. A magnetic tape that has been encoded with positioning information is installed along the elevator guide rail that runs the height of the elevator shaft, from top to bottom. A sensing device is attached to the elevator itself, which reads the encoded information and transmits it to a Controller Area Network (CAN) system. The combination of Schindler's MoSIS with CAN (originally developed by Robert Bosch GmbH for the auto industry) makes installation far faster, operation much safer and more precise, and maintenance much easier.

However, when Schindler looked at the potential cash return on the system, its executives did not like what they saw. As popular as MoSIS was sure to be, the acquisition of the knowledge required to create it involved a large investment, and Schindler believed it could not achieve payback solely through the sale of its own complete elevators equipped with the MoSIS system.

As a result of this analysis, Schindler might have chosen to license the MoSIS technology, including the magnetically encoded tape and the sensing unit, to other elevator manufacturers throughout the world. That way, Schindler would be able to achieve scale more quickly, reduce support costs, and gain payback faster. However, Schindler knew that the core technology of MoSIS could have a wide variety of applications in other industries and nonelevator product categories. By licensing, it might enable its competitors to gain a foothold in those other areas, which it was loath to do.

So Schindler made a clever move. It decided to license only part of the MoSIS system, the magnetic tape positioning technology but not the sensor unit. This would give Schindler a head start in selling its own elevators equipped with MoSIS, keep some of the intellectual assets out of the hands and minds of its potential competitors, and generate substantial license fees that could be applied to developing other applications for MoSIS.

In other words, Schindler wanted the licensing fees to help recoup the investment in knowledge acquisition, but did not want to risk the loss of future business in other industries by giving its licensees too much knowledge.

Enhancing the Brand

The licensor model can be brand enhancing if it creates or strengthens the perception that the company is innovative.

Through many means, including advertising and public relations, IBM regularly communicates to its constituencies about the revenue it receives from the licensing of its technologies. The company's goal is to raise the awareness of its technological prowess (which helps it attract and retain good employees) and its ability to turn intellectual assets into cash (which shareholders and potential investors always value). Licensing, and the awareness around it, has helped IBM burnish its image as an innovator.

Strengthening the Ecosystem

Licensing can also be used as a strategic move to gain an ecosystem benefit and preempt competitors.

Nokia made such a move in 1998, when Microsoft, the most successful licensor of our age, decided to expand its Windows franchise into mobile telephony. Microsoft's move had a dramatic effect on the other players in the mobile phone industry. They decided, in effect, to attempt to create a standard before Windows could get a foothold. Nokia, Sony Ericsson, Motorola, Psion, and Matsushita put together a consortium called Symbian to develop and license its own operating system software for mobile phones that would offer next-generation features and capabilities to trump Windows. This represented a major shift for Nokia, a company that had traditionally been an integrator. Nokia licensed its proprietary Series 60 user interface, along with the source code. Although Microsoft has made progress, many of the world's major mobile phone manufacturers—including LG Electronics, Lenovo,

Panasonic, Samsung, Sendo, and Siemens—also license the Series 60 operating system. All are Nokia competitors.

In essence, these companies, with Nokia leading the way, have come to a new view of innovation in the world of mobile phones. Rather than thinking of innovation as the "thing"—the tangible product that contains new technology features or capabilities—they have come to understand innovation as a framework of benefits and routes to payback. In this case, the need for ecosystem benefit catalyzed Nokia to choose the licensing model, which the company had long shunned, because it saw that the greatest economic value could eventually be gained through this indirect route.

Energizing the Organization

Licensing can be used to motivate and stimulate people within the organization, by providing another way for their ideas to be put to use and generate payback or indirect benefits for the company.

Motorola, for example, is using the licensor model to put some knowledge assets to work that might otherwise have gone unused. Motorola Labs, the company's applied research arm, developed a way to get carbon nanotubes (CNTs) onto the glass of large, flat-screen video displays by growing the CNTs directly on the glass. The breakthrough will enable the creation of flat-panel displays that will be brighter and have a wider viewing angle, and deliver excellent color quality, all in a superthin format that will be less expensive to produce than systems based on current technologies. Motorola dubbed the technology NanoEmissive Display.

Jim O'Connor, corporate vice president and general manager of the Early Stage Accelerator at Motorola, learned of the invention and decided to invest in its further development. "I authorized about $1 million of ESA funds for the project, but we have invested probably $200 million in this area over the last fifteen years," he said. O'Connor's investment funded the creation of a prototype, so Motorola and other manufacturers could assess the technology in a working model.

"What has traditionally happened in a lot of these inventions," O'Connor said, "is what happens in a lot of football games. The team gets to about the 5-yard line and then they just stop. They can't put the ball over the goal line, because that last bit is the hardest part."[9]

The prototype proved that the NanoEmissive Display had tremendous promise, but the question was whether and how to commercialize it. Motorola had once been in the TV manufacturing business, and O'Connor considered producing the new televisions as an integrator. But after reviewing the options, "We said, we're not going to build flat-panel displays," said O'Connor. "The best business model for the project is licensing. That's a great business. It's high-margin, low-cost."

Motorola is negotiating licenses with several of the major producers of television panel producers. But there also have already been noncash benefits to the company. "One is the brand," O'Connor said. "The fact that we developed this technology enhances the brand of Motorola as an innovator, which is a big deal because some people had forgotten that." There was also benefit to the Motorola organization. "It helps show people inside that, while we're creating stuff, we're actually announcing stuff. That helps builds a sense of confidence. 'Wow, we did that! We're really that technically good. We beat NEC. We beat Samsung.' It's a big thing for people to see their stuff come to life."

The Risks of Licensing

To gain maximum cash payback as a licensor requires intelligent and constant management of intellectual assets throughout the innovation process. Without such attention, the licensing effort can go seriously astray. The licensor faces a unique set of risks and challenges that must be weighed against the model's potentially attractive financial return.

A successful licensor must have very strong protection, create a standard, or stay at the leading edge of technology. Because the licensor is not involved in the commercialization or realization of the opportunity, it is unable to establish many of the other barriers that can preserve profits when a technological advantage disappears. Licensors must also work hard to stay in touch with end markets, so they have the knowledge to develop further ideas and technologies that customers will want and pay for. And licensors must be able to protect and aggressively defend their ideas. In fact, sometimes licensors must defend their ideas from the actions of their own licensees, by employing additional innovation business models. "Companies do not want one competitor to be able to license a technology and create a standard," Ad Huijser of Philips explained. "Even if they take a license, they will soon be trying to engineer around it to avoid paying a license fee where possible. This is why it is often critical for us to be active in manufacturing as well. You need a big commercial stick to keep the standard from becoming irrelevant."[10]

Despite these risks, a well-licensed technology, such as code division multiple access (CDMA) or the DVD, has extraordinary potential to generate payback for the company that creates it. That's why so many companies now insist on participating in any exercise that might result in a standard that could affect their industry. If an opportunity to license is going to be created, they want to be able to own a part of it.

Degussa, a specialty chemicals company, understands the risks of licensing. The company, which employs twenty-nine hundred people in some forty R&D facilities worldwide, creates a steady stream of new products for its sales force to commercialize. Typically, more than 20 percent of its products are less than five years old. Although Degussa usually integrates, it decided to use the licensor approach for a new product developed by Creavis, its technologies and innovation unit.

Why the change in approach? The new chemical was targeted at a single, highly specialized industry that Degussa had not participated in previously. When the company analyzed this new opportunity, it found that the potential customers had a much slower rate of adoption than in other industries. This meant that the cash curve of the new product would be much different than Degussa was used to, with larger start-up costs, a longer time to achieve scale, and higher support costs, all of which meant a longer journey to payback. "We decided we couldn't afford to invest the money and people required, given the other opportunities we had," said Andreas Gutsch, head of Creavis. "We decided to give this project to another company that was willing to take the license to make it, a company that was accustomed to the industry and could live with lower margins and slower sales growth."[11]

The licensing arrangement is not without risks. "Usually, if we put resources into something, and build up competencies on the technical and marketing side, we would like to materialize the business," said Gutsch. "A license on a patent does not necessarily guarantee success in the market. You need to have the market intelligence, the customer relations, and a huge variety of other things, to make it a success. If we give a license to a company that has no clue about the marketing strategy, for example, then it's not very likely they will make money out of the license. And, therefore, we will not either."

These risks and challenges of licensing can be, and are, overcome every day by innovative companies. The innovative licensor finds ways to make its product "sticky," ensuring payback long after patents have expired. Software companies, for instance, have long relied on the look and feel of the user interface to lock users into their product, even when more technologically advanced interfaces have become commercially available. That is precisely what TiVo, the digital video recorder developer, is banking on as the big consumer electronics firms increasingly offer similar technologies.

QUALCOMM: Smart Choice and Change of Model

QUALCOMM's experience shows both the opportunities for licensing and the skills needed to be successful in capitalizing on them. Founded in San Diego in 1985 by seven industry veterans, QUALCOMM developed and commercialized the CDMA technology that serves as the foundation for many wireless networks. In fiscal year 2005, nearly ninety percent of the company's sales and essentially all of its earnings come from two business segments: selling CDMA chipsets for mobile phones and collecting royalties on all equipment that uses CDMA technology. To support these businesses, QUALCOMM spends over $1 billion a year, or 18 percent of its revenue, on research and development.

Much of the QUALCOMM story centers on the perseverance of the chairman, former CEO, and one of the original developers of the CDMA technology, Dr. Irwin Jacobs, who had to convince a reluctant industry of the superiority of his technology. However, the company also presents an insightful case study on how to change innovation business models.

It took QUALCOMM years to win over customers. The first CDMA network was launched in Hong Kong in 1995, with the first network arriving in the United States in 1996.[12] Because manufacturers were slow to make the new CDMA phones, QUALCOMM began manufacturing them itself.[13] As Dr. Jacobs recalled in a 2003 interview, "We decided we'd better continue to develop the equipment commercially, so that when the standard was complete we could quickly launch the equipment. A late product could have killed it, because time was critical."[14] While QUALCOMM was able to shorten the time to market, it still had to commit a large postlaunch investment. However, once QUALCOMM got to market, rivals soon began offering CDMA phones, driving down prices and margins.

In 1999 QUALCOMM made a critical strategic model choice: to exit the hardware business and instead focus on chipsets for wireless devices and its intellectual property portfolio via a licensor strategy. At the time, Jacobs said, "QUALCOMM has always been based on technical innovation. Now we'll let others deal with wrapping plastic around chips."[15] However, the decision was not an easy one, and QUALCOMM executives had to weigh the risks and the rewards of such a move.

Because QUALCOMM's patents covered CDMA technology, the company was well positioned to make CDMA the industry standard for "third generation" wireless technologies. As Tim Luke, an analyst with Lehman Brothers, declared at the time, "They've got the core technology for the next millennium." Also, getting out of the low-margin manufacturing business meant that the company would have more funds for investments in research and development to maintain its technological leadership.[16] QUALCOMM shed its manufacturing businesses, selling its network equipment operations to Ericsson in 1998 and its handset manufacturing to Kyocera in 1999.[17]

However the risks of such a move were not insignificant. In the fast-changing world of communications technology, it's difficult for a company to keep its grip on proprietary IP rights.[18] The standards battle had not yet been definitely won, and it was not yet assured that QUALCOMM's technology would become the third-generation standard.[19] As the CEO of rival telecom chip maker Conexant Systems, Dwight Decker, said at the time, "QUALCOMM has fundamental knowledge of CDMA, but it's going to have to run real hard to keep up."[20] QUALCOMM also recognized the risks inherent in ceding control of the sales process and relying on other companies to sell the technology to customers.

The gamble paid off. CDMA did become the standard for third-generation wireless networks, and QUALCOMM's licensing strategy generated extraordinary payback. In 2005, QUALCOMM earnings were over $2 billion on sales of $5.7 billion, reflecting how a successful licensing approach can result in a cash curve that is extremely attractive—low ongoing investments, with most of the resulting revenue dropping to the bottom line.

But there are downsides, even to a licensing strategy as successful as QUALCOMM's. In particular, it can put a strain on ecosystem relationships. Licensees often see the fee as a "toll" and the licensor as a toll taker. "Our licensees don't like to pay us royalties, but they forget the work we put in to get the business," said Steve Altman, president of QUALCOMM.[21]

To improve ecosystem relationships, licensors often provide support and assistance to the licensees' commercialization efforts. This approach worked well for Ray Dolby and has also worked for QUALCOMM. "We provide the licensees with quite a bit more than a patent license," Altman said. "If we were just an IP shop, we would not be successful. What caused us to be a success was that very early on we didn't just license patents, we enabled the manufacturers to get to market quickly."[22]

An example of QUALCOMM's initiatives to help manufacturers get to market quickly is its BREW solution, which is written in C/C++, the programming language that is most popular with manufacturers. And it is an end-to-end solution for wireless applications development, which makes realization easier and more likely to generate payback.

The licensor strategy also requires that the company stay sharply focused on developments in technologies that relate to its business. As Altman put it, "QUALCOMM could have easily said, 'Let's close up shop, sit back and wait for royalties to come.' But that would have been a short-lived business: the technology evolves very quickly."[23]

Companies sometimes do not consider the licensing model, simply because they have misconceptions about its nature or because it is unfamiliar to them. It is not without risk and requires skilled execution. But by acknowledging that sometimes you are best served by having others commercialize and realize your idea and pay you for the privilege, you can generate a payback from licensing that can rival, even exceed, that of the other innovation business models.

PART THREE

Aligning and Leading for Payback

I N ORDER TO MAXIMIZE the payback from innovation, a company must align its organization around the innovation process and its leaders must actively, honestly, and consistently support it.

In chapter 7, Aligning, we focus on how the various organizational units, disciplines, activities, structures, and processes are aligned around driving innovation and creating payback. We explore the six factors that are essential to this alignment: individual responsibility, unit responsibility, companywide responsibility, conducive conditions, openness, and measurement.

In chapter 8, Leading, we learn how the leaders of some of the most successful innovators in the world go about their work. There are seven main areas where the innovation leader must engage and make key decisions: convincing an organization that innovation matters, allocating resources, choosing the innovation business model, focusing the organization on the right things, reshaping dynasties, putting the right people in the right place, and encouraging and modeling risk taking.

In the afterword, Taking Action, we offer some thoughts on what specific and practical steps to take to get started on the journey toward payback.

SEVEN

Aligning

One organization now owns and is accountable for innovations.

—Holger Schmidt, president, Special Systems Division,
   Siemens Medical Solutions

W HEN SENIOR EXECUTIVES ASK US, as they often do, "How can my company become more innovative?" we often discover that the question they're really asking is, "How can I increase the return on my innovation spending?" And often, they expect that the answer will have to do with organizational structure.

It would be much simpler if the secret of achieving payback were contained within a particular type of organizational structure and if that structure could be put in place, the innovation process would begin to operate more effectively. Of course, this is not true, and most businesspeople—even those who ask the question—know it's not.

We have found that companies can be innovative with almost any organizational structure—and can fail to be innovative with any structure, as well. What matters is alignment. Are the various units, disciplines, activities, structures, and processes aligned around driving innovation and creating payback? Or is the organization a patchwork, with some groups and aspects supporting and promoting those objectives, some blocking them, and some plodding along on another path altogether?

So although there is no single organizational structure that is the most innovation friendly, organizational alignment is fundamentally important. Alignment, or lack of it, affects every aspect of innovation, including the idea generation and evaluation process, the shape of the cash curve, the functioning of the business model, and the ability of the company to achieve payback.

Unfortunately, the effect of any organization on innovation is often a negative one. This is because organizations, no matter how untraditional they may be, are primarily designed for control, standardization, and reduction of risk—and these characteristics can be the enemies of innovation. That's what many executives told us in the 2006 BCG/ *BusinessWeek* "Senior Management Survey on Innovation," saying things like "We have too much centralization, or not enough, or an uneasy combination of both," and "We have too much group decision making," or "Insufficient delegation of responsibilities and accountability," or "We don't focus enough on payback," and, a common one, "Internal departmental silos make innovation and entrepreneurship difficult."

But innovative companies, of whatever size and shape, find ways to overcome these problems. In our survey with *BusinessWeek*, the top-ranked companies were, in descending order, Apple, Google, 3M, Toyota, Microsoft, GE, P&G, Nokia, Starbucks, and IBM.

These companies have very different organizational structures and cultures, yet all of them are aligned around innovation and achieve payback. Conversely, companies that have had limited success with innovation are not aligned—and usually the lack of alignment is due to one, or more, of the following factors:

- **Innovation strategy is at odds with business strategy.** The
  company's fundamental business strategy—its plan for achieving and building competitive advantage—is not supported by
  its innovation strategy. The spending on innovation is too
  high, too low, or focused on the wrong areas. Result: ideas and
  inventions are generated that the company doesn't want or
  can't effectively sell. The investment in resources doesn't advance the company's overall position and doesn't create sufficient payback.

- **Innovation is all talk and no support.** The leader says that he
  wants the company to be more innovative. But he invests very
  little in innovation capability and capacity. The research, development, engineering and manufacturing groups focus on
  more immediate concerns, such as cost cutting and quality improvement, while the sales organization puts all its weight behind selling current products and services and doesn't support
  anything new. The company's customers are poorly understood. Results: the organization doesn't know how to go about
  "being more innovative"; the leader becomes frustrated.

- **Innovation is an island.** A small innovation group is formed to
  generate new ideas and propose new products. It is headed by
  a leader who does not have enough authority or respect from
  the operating groups, or whose efforts are stymied by metrics
  that are inappropriate for the effort or cannot be achieved.
  R&D is focused on creating improvements and additions to
  existing successful products. Engineering and manufacturing
  are focused on expanding production. Result: the innovation
  group generates lots of ideas, some that gain internal investment, but few of which are realized.

- **The process is fragmented.** The leader does not spend time on
  innovation. R&D does not have much connection with customers.

Sales is focused on quarterly numbers. Support of products stops shortly after launch. Result: new products are launched, wither, and often quickly die.

- **Dynasties monopolize innovation resources.** The leader supports innovation. One product area or brand accounts for a large percentage of sales and profits and becomes a dynasty. There is no independent group whose charge is to foster innovation outside the dynasty; or if there is one, it does not have sufficient support or clout to take any real action. The leader does not reallocate resources away from the dynasty. Result: new ideas don't get funded or commercialized. Supporting new ideas is criticized as "milking the business."

- **Metrics confound the goals of innovation.** The leader supports innovation. Innovation capability is available. But key metrics get in the way. For example, production is measured on cost cutting. Salespeople are measured only on volume. Result: new products don't get enough attention, leading to poor quality, late market entry, or too little postlaunch investment.

All of these ways to be "unaligned" can have a negative effect on a company's ability to achieve payback. The opposite is also true.

Citigroup: Alignment as a Series of Small Steps

The story of the emerging markets business of Citigroup shows how many organizational alignment issues—some of them very small—are essential to creating alignment around innovation.

In the summer of 2000, Victor Menezes was tapped to lead Citigroup's emerging markets business and charged with realizing the group's potential to be the company's growth engine. "By creating a single management structure for these emerging markets," read a corporate statement, "the company is aligning its operations to respond to the growth potential and to the customer needs within these regions."[1] Menezes decided to focus on innovation as the best way to generate growth across the ninety countries that composed Citigroup's emerging markets business—an operation that employed some 50,000 staff members and generated $10–11 billion in annual revenue.

An innovation team was formed, headed by Claus Friis. "We first did a thorough assessment of the existing state of innovation in the business," Friis told us. "And based on that assessment, we came to three main conclusions. We had plenty of ideas for new products and services, but lacked the disciplined execution needed to achieve the large payback potential. The decision-making process was bureaucratic, and no one person had responsibility for innovation. And we found it difficult to replicate an innovation success from one country to another."[2]

Taking the first step in solving these problems, Friis and his task force created a new position, called innovation catalyst (IC), which would have the responsibilities of chief innovator (the main driver of innovation) rather than innovation facilitator (educator and mentor). Each IC would be responsible for the entire innovation process in a particular country or region and would have organizational authority, while accountability remained with the line managers who owned the P&L.

Candidates for the IC positions needed to have a number of characteristics, including a strong reputation within the region or country and enough seniority to have authority to make changes, a thorough understanding of the business and experience across products and functions, a collaborative approach and good facilitation skills, and a strong sense of both urgency and discipline.

The ICs were selected, attended a weeklong training session, and set to work. One of their tasks was to develop an innovation blueprint that showed what the process should look like, but that the countries were free to adapt and modify as appropriate. Not surprisingly, as the programs were implemented, it became clear there was a correlation between the quality of the IC and the organization's ability to innovate successfully.

Elcio Pereira, for example, was the IC for the Latin America region. As he began to develop the innovation effort, he discovered that a similar initiative had been launched in Brazil in 1994. There, a team had used reengineering tools to map the innovation process and had identified seventy gaps, conflicts, and disconnects. The Brazilian team had revamped the process, and it soon began producing results. Although there had been efforts to replicate the process in other countries, there was not sufficient support from the country leadership, and the scaling up of the success fizzled. (The innovation process remained in effect in Brazil, however, and is still producing positive results.)[3]

Pereira did not want the same thing to happen again, so he gathered the top fifty leaders for Citigroup Latin America to discuss the rollout of the new process. By the end of the meeting, he had secured their commitment and buy-in, and they had agreed on a number of goals and measures. In particular, each country committed to more than double the "innovation index"—the percentage of revenue in the last twelve months from products and services brought to market within the past two years—from its current 7 percent to 15 percent. They agreed to work toward a better balance between new "one-time" products (such as a one-time investment banking transaction) and new annuity revenues. They also agreed to work to reduce time to market, by improving their decision-making process, shortening the time to volume, and doing a better job of sharing knowledge about successful innovations.[4]

The innovation task force branded the initiative at Citigroup as i2i—meaning idea to implementation—and focused their efforts on four main areas: idea generation, decisions and approvals, implementation, and replication.[5]

To improve their ability to generate ideas, i2i created a "customer intimacy toolkit" that included a number of programs and features. The Customer of the Month program centered around a full-day meeting with a team representing a corporate customer, with the goal of identifying customer needs and brainstorming solutions. An Industry Needs Group identified needs and developed possible solutions for an entire industry; representatives from customers in that industry always participated in these sessions. A Customer Ecosystem tool enabled the Citigroup team to analyze a customer's entire ecosystem (e.g., their relationships with suppliers, the government, and others) to uncover revenue opportunities.[6]

The IC for each country also served as the first point of submission for new ideas. In the past, employees had been required to make a formal presentation of their new ideas to a review committee. The committee had the power to reject a proposal on the spot, and the "no" was often pronounced in front of a large group of people. The prospect of rejection by the committee could have a chilling effect on people with worthy ideas, and many accordingly chose not to bring their ideas forward. With the new process, however, the person with an idea could discuss it informally with the IC. The IC could check the company's innovation database (a centralized repository for all new ideas) to ensure that the idea was in fact new and different enough to pursue. If it was, the IC would help the idea generator refine and modify the proposal. The IC also talked with customers to get their reactions to the idea and see whether there was a real need for the proposed invention.[7]

After validating the idea, the IC then submitted it to the Magnet Team (MT) for approval. The MT was a dedicated, senior-level cross-functional team composed of representatives from product areas, legal, risk, tax, accounting, operations, technology, and compliance, with all members having equal authority. The MT members had to be available every day between 4:30 and 6:00 p.m., in case the IC called a meeting for idea review. Each MT member had to designate a colleague to attend the meeting if the member was not available. In this way, the MT acted as a one-stop approval process by ensuring senior-level representation from across the organization and by always being available at a regular time.[8]

A key improvement of the MT process was that it minimized rework and the requirement to shop around an idea for approval. The idea generator only had to present the idea once to the MT, which would immediately ask questions and raise concerns. All MT decisions were documented on the spot and circulated to senior management for transparency. There was also a comprehensive list of triggers that indicated the ideas that would need to be approved by senior executives at Citigroup headquarters in New York. The MT coordinated the interactions until the final decision was made.[9]

Once an idea was approved by the MT, it was passed along to a Deal Team (DT) for implementation. DTs consisted of empowered cross-functional members and were led by a Deal Champion, who was appointed by the MT. The DT was responsible for full implementation of approved ideas, according to the terms and conditions of the MT approval.[10]

Several tools were put in place to encourage knowledge transfer of successful innovations in countries and regions where market conditions were similar and the idea might prove relevant. Before this innovation initiative was launched, Citigroup had often watched as its successful innovations in one country were replicated in other markets by its competitors before Citigroup could do so itself.[11] To counter this, Pereira used product forums and customer solutions meetings to encourage new product transfer. Product managers were also required to identify places where successes could be relevant in order to ensure targeted transferring of good ideas. The knowledge system was used to capture and document all successes, thus providing a central repository of successful ideas for all of Latin America to tap into.[12]

Measuring innovation performance was also a key part of the program. While each country's innovation index was the primary metric used to assess progress, other metrics were used, including cycle time (from idea to launch or first transaction), number of ideas generated, number of ideas approved, conversion ratio (number of ideas approved to number of ideas generated), innovation revenues by product, and innovation revenues by customer.[13]

Rewards and incentives were created to reinforce the new innovation behaviors. Each quarter, two Innovation Awards were given out in Latin America, as well as two Replication Awards. The criteria for each award were based on payback, teamwork, speed, and managing complexity. To help celebrate the awards, quarterly Web ceremonies—led by the CEO of Latin America Corporate and Investment Banking—were held to present the awards and increase exposure to the successes across all of Latin America. Separate from the awards, a point system was created so that points were given to team members and individuals for ideas suggested, ideas chosen, and ideas successfully implemented. The points were redeemable for various gifts and travel awards. By using both frequent smaller awards and quarterly awards that were publicized across Latin America, Citigroup was able to generate enthusiasm for the innovation program.[14]

As a result of Citigroup's work in aligning the innovation process, its emerging markets business became more able to propose and act on new ideas and to replicate innovation practices across geographies, with the result that the business unit was able to increase its region-wide revenues from new products and services from about 7 percent of the total when the initiative began to 16 percent just two years later, surpassing the 15 percent target. In some Latin American countries where the innovation process was personally led by the country CEO, the innovation index surpassed thirty percent.

How Companies Align

Alignment looks different in every company, and is, accordingly, a state of being that is almost as difficult to describe as it is to execute. Alignment means that a company's business strategy, innovation process, innovation business model, organization, and leadership approach all are essentially geared toward and supportive of the innovation process and the company's payback goals. (It is, of course, possible to align the organization around a different activity, such as M&A or cost cutting, but don't expect such an alignment to result in innovation!)

Although this is not a comprehensive list, the most important elements that innovative companies pay attention to include:

- Individual responsibility

- Unit responsibility

- Companywide responsibility

- Conducive conditions

- Openness

- Measurement

Individual Responsibility

Innovation has to be someone's main operational responsibility. There needs to be a person who wakes up each morning worrying about how to execute a required set of innovation-related tasks that day and how to get the people in the organization to accomplish them. That person must be held accountable for the company's payback goals and measured on the ability of the company to achieve them.

In most organizations there is no such position. However, if you ask the CEO or president who owns the innovation process and is responsible for ensuring that payback is achieved, as we often do, they rarely answer "Me." Some of them acknowledge that they have de facto responsibility for innovation but, even so, will readily admit that they are not actively engaged in managing the process and results.

A few companies do have a "head of innovation" position, with a title like vice president of innovation or chief innovation officer. But too often the job comes with little authority, few resources, and not much operational support from the leader or the organization as a whole. As a result, the person and the organization are quickly marginalized and can even become a deterrent to innovation.

Whatever the title, there are two main ways to define the position with responsibility for innovation: the innovation facilitator or the chief innovator. They operate in different ways and serve different functions.

Innovation facilitator. The innovation facilitator is essentially an educator, advocate, and adviser. The facilitator usually has a team, and its responsibilities are to make the organization aware of the importance of innovation, standardize and communicate a vocabulary of innovation, provide a set of tools, develop training activities, and determine which metrics will be used. Usually, the business units determine whether and how to best use what the innovation facilitator has developed, often in consultation with, and support from, the leader.

The innovation facilitator is an especially important role for companies that are making a fundamental change from some other strategic alignment, because people in the organization will have to change attitudes and learn new skills and behaviors. Education, execution, measurement, and communication tools and approaches will have to be created from scratch, and the innovation facilitator may have to be involved for the many years such a change process can take. Ideally, facilitators eventually work themselves out of a job (or the job changes significantly), because the ways of innovation become thoroughly embedded in the organization, and constant, intensive, companywide education is no longer required.

Chief innovator. The chief innovator is responsible for managing the entire innovation process holistically and makes decisions about the trade-offs involved. Chief innovators don't do everything themselves, but neither do they allow organizations to not perform. They clear bottlenecks when it becomes necessary to do so, get involved in both the strategic and the tactical, and have either the authority or the personal influence to make things happen.

Chief innovator is a role often played by the CEO or chairman, especially when the company must make a change from another business strategy, such as cost cutting, merger and acquisition, or geographic expansion. When that is the case, the CEO is often the only person who can make the decision to change the alignment from the company's current focus and move toward innovation.

Steve Jobs is a chief innovator, deeply involved in making sure ideas are effectively and efficiently turned into offerings customers and consumers can buy. Another chief executive who is also chief innovator is Doh-Seok Choi, president and chief financial officer of Samsung.

Just a little more than a decade ago, Samsung was known as a me-too producer of low- and medium-quality electronics goods. The Samsung name was not really considered a brand; it was seen as just another Asian nameplate. To move from a cost-cutting strategy to innovation has required the intense involvement of President Choi, playing the role of chief innovator. He is also chief financial officer; at Samsung, innovation is always pursued for payback.

Companies that don't have a chief innovator, or whose chief innovator does not have visibility into the entire process and the ability to manage it, can get in trouble.

At one large maker of consumer products, for example, the CEO set his company on a course toward innovation. Although he did not play the role of chief innovator, he did get the innovation ball rolling. He focused primarily on idea generation and commercialization, and the company became adept at generating and evaluating ideas for new products, quickly commercializing them, and creating capacity to produce them at volume. However, there was a disconnect between commercialization and realization. The sales force and distribution channel had not fully participated in the innovation process and did not have the resources or motivation to sell and support the avalanche of new products that burst on them, seemingly without warning, and threatened to overwhelm them.

Naturally, the sales force focused their efforts on the handful of products they understood, believed in, and could effectively support. The remaining products—which had been developed at great expense to the company and many of which had strong potential for payback— were essentially thrown onto the market without support. Not surprisingly, they failed to live up to expectations and were even seen by the company as failures. Not only was payback negatively affected, people in the organization were frustrated and began blaming each other for the lack of return. Had innovation been someone's main responsibility, the gap in the process might have been observed and bridged, the infighting and recriminations could have been avoided, and payback could have been achieved. Hundreds of millions of dollars of shareholder value that could have been created, was not—because of the gap between commercialization and realization.

Unit Responsibility

Innovative companies often establish small groups or discrete units that encourage or support innovation in specific ways. Some are innovation incubators, designed to encourage, seek out, evaluate, and promote a wide variety of ideas and inventions. Some focus on the creation of a single new product or service, and often operate outside the normal operations of the company to do so, like a skunk works. Others function like internal venture capitalists or sponsors, selecting and funding ideas and pushing them through to commercialization. Others play different roles. Most focus on ideas and opportunities that have potential for significant payback but require an effort that existing business units are unwilling, or unable, to pursue.

The responsibility of Motorola's Early Stage Accelerator (ESA), for example, is to review internal ideas for new products and services, investigate their potential for payback, and accelerate the incubation of those with high potential to determine their validity. Sometimes the ESA will find a good idea that had been shelved and resurrect it. Sometimes it will support a promising invention that has been unable to get attention within the organization. Once the ESA identifies a promising opportunity, it completes rapid prototypes, conducts pilot and market research, and formulates a business plan, including which innovation business model to use.

"We accelerate, catalyze, commercialize," Jim O'Connor of Motorola explained. "We're like a middleware layer that says, 'OK, guys, we can make that into a product.' We take ideas, technologies, and concepts that would otherwise sit there, and turn them into something we can make money with."[15] The role is essential, O'Connor said, because the company's business units are focused on current customers and operating earnings targets, and put most of their efforts into product extensions or incremental improvements. The group has supported more than thirty projects, six of which have "graduated" into one of Motorola's main businesses.

Companywide Responsibility

While innovation must be someone's operational responsibility, supported appropriately, it must also be everyone's job. Ideas, of any size or application, can in fact come from everywhere. Senior executives or researchers in a corporation have no more likelihood of developing an idea or insight that will result in a product or process innovation and financial returns than an interested employee on the shop floor or someone working with customers every day.

At Samsung, for example, all employees, not just product designers or engineers, think of innovation as central to their role. The willingness to change and keep changing, to think and rethink, and to innovate was almost tangible in everyone we met at the Gihueng complex in Seoul, South Korea. One employee put it best when he said, "We believe that everyone can make a contribution to innovation at Samsung, not just people developing products. What we have accomplished here has allowed us to beat our competitors to market and further improve our position."

The Gihueng complex, which employs twenty-five thousand people, demonstrates how the company has pushed innovation into all parts of its business and made it part of everyone's job. The semiconductor business's rallying cry is "Creating the future with a nomad spirit." Every employee is called upon to be a team player, a pioneer, and an innovator. Innovation, which in Samsung must result in an impact on the bottom line, shows up everywhere.

One way to be sure that innovation is everyone's job—and that everyone understands that job in a consistent way—is to be very precise about the language of innovation and how it is used throughout the company. At Samsung, President Choi told us, "The first answer to how we are aligned would be that the language used by all the employees—the keywords—is unified. The chairman sets the strategy and everyone in the company understands it clearly. Everyone uses the same language. This is a big change. Chairman Lee said, 'You must be prepared to change everything about yourself except your wife and children to be successful.' Unifying the language and sharing the same thoughts is obviously not easy—it's a long, arduous process. We are still at sea."[16]

We heard these same messages communicated consistently everywhere we went in the company, and most used exactly the same words.

Conducive Conditions

Although we've said that a lack of ideas is rarely an issue for most companies, it is an issue for some—and no company has too many really good ideas with great potential for payback. However, many companies have, over the past fifteen years, taken steps that lessen the likelihood that people inside the organization will be able to come up with original insights and ideas that have potential for generating large returns.

Leaders of innovative companies realize how important it is to create conditions that support innovation and encourage creativity. As Jean-Louis Ricaud, executive vice president of engineering and quality at Renault, put it, "I think we need to protect the 'vision' people, to create a condition that allows them to be creative, and we also have to be able to challenge them and to coax them."[17]

Six conditions can help people be more effective in the innovation process, particularly the idea generation phase, no matter what the formal organization structure may be:

- Time to think

- Space to explore

- Deep domain knowledge

- Stimulation

- A challenging environment

- Motivation

Time to think. Many companies have eliminated positions that dealt with issues and ideas that were unrelated to immediate output of products or services. People who might have had time to think in the past have often been required to take on new tasks and more responsibilities, and their thinking time has been reduced or obliterated.

Most executives will admit that they don't have enough time to think about much beyond their current activities. "You can't think about something new when you have to spend all your time taking care of the million things you have on your plate right now," an executive told us. Our studies suggest that most white-collar individuals in the corporate world today spend far less than 5 percent of their time on activities they believe are about generating new ideas or insights.

Innovative companies allow—even encourage—their people to spend some of their time thinking about new things. At Robert Bosch GmbH, "to generate ideas in advanced research and development, department managers and group leaders can spend at least 5 percent of their people's capacity to open up new ideas without arguing about what they want to spend this capacity on. We do this because if you have an idea and ten minutes later you have to argue why it is a good idea, it is very simple to kill it," Dr. Siegfried Dais, deputy chairman of the board of management at Bosch, told us. "Later, they have to defend their investments because there is no free lunch. But not in advance."[18]

Space to explore. People need both psychic and physical space—the freedom to think and dream as well as some walking-around space in which those activities are expected, even protected. When people are part of an environment where thinking is encouraged and where "crazy" ideas are tolerated, they feel they are being given permission to think about new, even risky, ideas and approaches.

One of our clients, for example, has seen that certain geographic organizations (such as Australia and South Africa) have become hotbeds of innovation. He believes this is because the talented senior managers in those geographies, who operate outside of the control of the corporate center, have the space they need to generate lots of good ideas. They also have the full set of functions and capabilities required to realize and commercialize the ideas, which would not be the case if innovation were more centralized.

Deep domain knowledge. Another requirement for innovation is to possess or obtain a great deal of knowledge about an issue, problem, or inquiry; it's virtually impossible to generate innovative ideas about something you know little or nothing about. Although breakthrough idea-generators are often outsiders to the discipline to which they come with a great idea, they usually possess deep domain knowledge of some kind and with some relevance, however tangential, to their idea.

Shuji Nakamura, who was instrumental in the development of a key technology that led to the creation of the blue light-emitting diode (LED) and the blue semiconductor laser, provides an example of how deep domain knowledge can pay off in the generation of new ideas. After graduating with a Master in Electrical Engineering in 1979 from the University of Tokushima in Japan, Nakamura joined Nichia Corporation as a researcher. He and a team of colleagues and assistants, well-financed and with their own research facility, spent many years building their knowledge about LED technology. In 1988, the company sent Dr. Nakamura to study at the University of Florida, where he gained further understanding of related technologies. In 1993, Nichia announced that Nakamura had succeeded in fabricating a bright blue light-emitting diode. "I'd call it a piece of technical work that is absolutely remarkable, and stands on its own,"[19] said Bill Lenth, manager of science and technology at IBM.

The blue LED facilitated the development of more-vibrant LED displays and also led to the development of the white LED, which may eventually replace traditional light bulbs.[20] By leveraging the discoveries made by Nakamura and his team, and incorporating the work of many other engineers and scientists, Nichia was able to take another step forward—the development of the blue-violet semiconductor laser, which the company announced in 1999. It provides the technical basis for a new wave of media and data storage products, such as the Blu-ray Disc and HD-DVD. Nakamura's deep domain knowledge was acquired through formal education, corporate research, team collaboration, and experimentation. Without it, Nakamura could never have "stumbled across" the technology that made the blue LED possible or even have been credible enough to assert that such a thing could exist and be encouraged by his company to realize it.

The ability to develop deep domain knowledge has been hampered by the same corporate actions that have put the squeeze on the time and space available for idea generation. In particular, companies that have reduced spending on education and training often find that their domain knowledge is not as deep as it once was, and sometimes not as deep as that of its competitors or partners. In addition, the loss of people whose primary function is to think and explore, along with rapid job rotation, has further depleted the store of knowledge at many companies.

To make matters worse, many countries (including Japan, the United States, many European nations and Canada) are facing an imminent knowledge crisis as an entire generation of experienced workers retires over the next several years. With them will go an extraordinary amount of expertise and knowledge, the fuel of idea generation.

Stimulation. Even with time and space to think, and plenty of domain knowledge to think about, people generally come up with more and better ideas when they receive some stimulation in their thinking from the influence of other people and from exposure to ideas that are not their own. "New ideas or unique ideas come from conflicts with people from different cultures or different ways of thinking," Teruaki Aoki, senior executive vice president and executive officer of Sony, told us. "It is not easy to manage people from different cultures, but once you can manage them, I think you can expect something new to come up."[21]

A study by Karim R. Lakhani, one of our colleagues and an assistant professor at Harvard Business School, shows the importance of external sources of innovation to a company's innovation efforts. Lakhani and his colleagues studied InnoCentive.com, which broadcasts unsolved scientific problems of *Fortune* 1000 firms to InnoCentive.com's network of 90,000 scientists. The scientists select the problems that interest them and develop solutions to them in hopes of winning an award. Lakhani and his colleagues found that 73 percent of those who were successful at solving the problems they had selected said that their solutions were to some degree based on ideas they had already developed themselves or on ideas they already knew about that had been generated by someone else.[22]

According to Lakhani, there was a statistically significant correlation between a scientist's solving a problem successfully and the scientist's indication that the problem was outside his field of expertise or at the boundary of it. Essentially, the external scientists were bridging disciplines, by taking knowledge from somewhere else, looking at a problem with fresh eyes and a new perspective, and saying, "We've already solved this or something like this in our field."

We have talked with countless executives who say that the best new ideas, especially those that solve a long-standing problem, are often catalyzed by just this kind of "fresh perspective." Innovative companies find ways to bring new influences into the workplace and expose their people to outside ideas. A trip to the museum. The distribution of a provocative book. A speaker series. Erwin Schrodinger's public lectures on the nature of the gene and potential relationship between physics and biology, for example, are credited with encouraging physicists to study biology and vice versa, resulting in significant contributions toward the initiation of molecular biology.[23]

A challenging environment. Ideas get generated, tested, refined, and improved in environments where ideas are valued and constantly focused on—as opposed to environments where new ideas don't get aggressively challenged and refined. In fact, asking the right questions about an idea is key to developing ideas that will stand the test of the market-place. Dr. Siegfried Dais at Bosch put it this way: "The most complex thing is identifying the relevant questions, those areas which will determine your success five years from now. That's the most challenging and demanding task of those in charge of innovation or future success—making sure we are asking the right questions."[24]

In highly innovative companies, ideas are constantly being proposed, talked about, aggressively challenged, tested, poked, and prodded. Nothing is left unquestioned. Nothing is disregarded. In such environments, challenges and questions must fly from everywhere and on every aspect of the idea. "Why?" is the most frequently heard question, along with "Why not?" "What will it take to make this idea work?" and "What value would this have to customers?"

LG Electronics (LGE) has a well-developed innovation methodology that ensures that the challenging of ideas will be part of the innovation process. The methodology centers on Tear Down and Redesign (TDR), an activity in which an idea or product concept is completely torn apart and rebuilt from the ground up. The TDR process was created in 1995 by S. S. Kim, CEO of LGE, in its Digital Appliance Company. TDR has six main elements: cross-functional teams, strict time limits, a focus on big issues, stretch performance goals ("5 percent may be impossible, but 30 percent is possible"), full-time participation by team members, and dedicated TDR physical locations where the entire team works together. (There are TDR rooms in each of LGE's major facilities.)[25]

We visited a TDR room in the Chang Wong Complex of LGE's Digital Appliance Company. To enter it, the visitor must proceed through several security stops; authorized employees are recognized by a retina scan. As many as ten teams of eight to fifteen people work in the TDR. Members selected for each team know it is a great honor. They also know that they will be stretched to the limit, they are expected to perform, and the time they spend at the TDR will likely be the most challenging of their careers.

The teams we observed were working intensely, preparing for their regular report to the president of their division—and the president of the entire Domestic Appliances division may also attend. They know that the meeting will not be a perfunctory review. The president is sure to ask significant and difficult questions. If he is not satisfied with the answers, the project may very well be canceled on the spot. And the president is always well prepared. Each month, he receives a full report on each team detailing its progress in three major categories: the expected financial performance of the innovation, project-specific key performance indicators (KPIs), and new knowledge being created.

President Lee of the Domestic Appliances division told us that they had some 200 projects underway, and he would personally review fifty or sixty of them at a two-day meeting held each month. (He has held 106 of these sessions in the past ten years.) Generally, half of the projects will have some impact on the business's overall strategy. Ten to twenty of them are "royal" projects—ones that are expected to have a major impact on the business and therefore receive intense attention, greater investment than the others, and strong personal support from President Lee. His involvement is not unusual—throughout LGE, the TDR process receives engagement from senior executives, is aligned with employee incentives and rewards, and is a constant focus of communications throughout the organization.[26]

Motivation. Finally, the process of innovation requires that people be motivated to come up with ideas. Many people are self-motivated to do so; they have a natural desire to do something that hasn't been done before or to answer a question that intrigues them. But many people need a nudge.

Motivation is particularly important because new ideas generally imply that the organization will be asked to do something it has never done before. This will always mean that there will be hurdles to leap and barriers to be removed. Only a sufficiently motivated person will have the energy and commitment to press forward in the face of adversity. The creation of something new also may require working long and extra hours, coming up with endless iterations of a single idea, and persevering through failures and unexpected difficulties. Even when the motivation to generate and pursue new ideas is intrinsic, extrinsic rewards—including financial compensation, recognition, and access to new opportunities—can help.

One of the reasons Sony developed its Qualia line of high-end consumer electronics was to motivate its own people. The Qualia line is composed of home entertainment electronics products of exceptional performance and design that sell at high price points (e.g., $11,000 for a 36-inch TV set, $7,000 for a CD player, $2,600 for a pair of headphones). But they are not meant to be exclusive products, marketed only to wealthy customers. They were meant for people who care deeply about the experience of home entertainment and want the very best products available. "Some people think that Sony's Qualia products are for a small group of wealthy people, but this is not right," said Idei Nobuyuki, Sony's former chairman. "They are not about a simple economic logic."[27]

Makoto Kogure, senior vice president of Sony and president of Sony TV, explained the rationale to us in a conversation at his office: "TVs are becoming a commodity product. At the same time, our customers expect something different from Sony. We have to show them where we are going and make sure they understand what we can do. But we had gotten to a point internally where our TV engineers, who had been in the CRT business for so long, were very focused on costs, given how competitive the market was. Perhaps sometimes they forgot the very high standards that we were built on. So I wanted to get them to focus on the very best—the best resolution, the best design —something that wasn't a commodity. That's of course a very hard change. Some people knew how to do it, some didn't." How did he change the mind-set? Kogure was matter-of-fact. "I told them to spend as much money as they needed to. The amount of money we will make on these products is small, but I thought it was important for our engineers and their excitement to make sure that the best TVs in the world were Sonys.

"When I first told the engineers to do this," Kogure said, "they didn't know what to do. But they quickly figured out how to make the best."[28]

Openness

A company and its leaders can do a great deal to foster and manage innovation within its own walls, but, increasingly, they seek to take advantage of knowledge and expertise available outside the organization.

Although the subject of "open innovation" has gotten a great deal of attention, most companies are still relying on a few traditional external sources for ideas—such as university engineering departments and customers—and have not begun to look more broadly.

According to BCG's proprietary innovation benchmarking database, half of all executives feel that their companies have a good understanding of what customers want and what the competition is doing, and also of what new technologies are emerging from academia and related industries. But when it comes to using that external understanding, less than a quarter of respondents believe that they are effectively upgrading their innovation knowledge via partners, suppliers, and academia. In other words, executives understand the need to scan what's going on around them, yet they don't know how to successfully process that knowledge for use toward their own innovation success.

The question most companies wrestle with is not whether to tap into external sources but how, and how to share any resulting rewards. How specifically do you overcome both the tactical and the organizational barriers that exist to moving from a good idea (in this case, tapping into the outside world) to actually making it happen? Companies that are serious about finding ideas outside their boundaries follow one of two basic models: the scout or the beacon.

The scout. The scout is the company that is perpetually going outside and looking for ideas. There are several requirements to being an effective scout. The most important is that scouts need to know what it is they are looking for. It is difficult to be effective if you are merely searching for "a good idea." Generally, the most effective scouts have a well-defined understanding of the specific areas they are seeking new ideas and innovations from. The second requirement is that they need to be able to find the areas or people that they are interested in.

Schindler, the elevator maker, is a good example of a scout company. Karl Weinberger, head of corporate R&D at Schindler, told us, "We are not a basic research company. Very rarely do we approve a research budget that has to start at zero. Instead, we are very good scouts. We have a separate technology management group, and one of its major tasks is to find the best technology available in the world and see how we can adapt these technologies in our application and to our needs. That's the secret of these successes. We don't necessarily have to come to these ideas ourselves, but discovering what other people were doing gave us an accelerant to our thinking. And when we did it right, we get a great payback on our investment in finding the knowledge."[29]

In one of its scouting forays, Schindler came across an interesting material that it had not known about before and that had properties that looked promising for elevator applications. One of the most important of an elevator's systems is the safety brake, which in an emergency clamps onto a guide rail to bring the car to a stop. An enormous amount of energy is released when the brake is deployed. According to Schindler's calculations, the lining of a brake that is used in a free-falling elevator car can heat to 1,200 degrees Celsius. Conventional steel brake linings are unable to withstand such heat. But Schindler found a material that could, in the ceramic tiles used to protect the space shuttle as it reenters the earth's atmosphere. High-performance ceramic is extremely resistant to heat, is durable, and produces a high braking effect. Schindler identified the opportunity through work with the German space agency *Deutsches Zentrum für Luft und Raumfahrt* (DLR), and developed a new elevator braking system that improved performance by more than 33 percent, with a 65 percent reduction in volume, a 35 percent reduction in weight, and a 10 percent reduction in the number of parts.

The beacon. The beacon is the company that has made enough of a name for itself in its area that inventors, technologists, other companies with ideas, or collaborators of potential interest seek out that company because they know what they are interested in and where to go.

Some companies, after successfully experimenting with scouting, realize that they will never be able to reach out to enough of the known sources of ideas of relevance and do not have the capability to identify the currently unknown ones. Accordingly, they begin to move from being a scout to being a beacon. They let it be known that they want people—wherever they are in the world and in whatever discipline— to bring ideas to them, that the ideas will be welcomed, and that a well-defined process will be in place to receive and manage them.

Procter & Gamble has invested significant effort in developing not only its scouting capabilities but also its ability to be a beacon. Jeff Weedman, vice president of external business development at P&G, puts it this way: "We want to be the place where people come first with their ideas. They'll do that because they know what we're looking for, they know we'll respect them and their ideas, and they know that we'll work with them to commercialize and realize the idea better than anybody else. And if the idea doesn't fit our corporate strategy and we can't use it, we probably know who can and we'll pass it along to them, even if it's a competitor."[30]

Measurement

We once asked a European financial services executive, "How do you measure the performance of your innovation efforts? Please be as specific as possible."

He replied, with classic bluntness, "We don't."

Companies use many measures to track their innovation performance. Three of the most popular are the percentage of sales that is generated by new products (usually defined as less than three years old), the number of patents the company files in a given year, and overall revenue growth. These and a multitude of other metrics are useful, but innovation is so broad and complex that no single measure can track innovation performance. And none of these has much to do with cash payback.

Four aspects of the innovation process need to be measured:[31]

- The inputs to the process

- The performance of the process

- The cash payback

- The indirect benefits

Inputs. Inputs mean resources, such as money and people. Some inputs will be in short supply and this scarcity can become a bottleneck. Other inputs are readily available and this scarcity can be scaled up and down with relative ease. The most important inputs will vary depending on the situation, but most management teams track some of the following items:

- The number of ideas, with expected payback potential for each

- The number of full-time equivalent staff for selected functions involved in the process—and, most importantly, what their key people are working on

- The operating expense

- The capital expenditure

Performance. The inputs are acted on by certain people and processes, which can be tracked and measured, including:

- Cycle time through the entire innovation process

- Cycle time through specific parts of the process (e.g., prototype or pilot run)

- Deviation around average cycle times—high end to low end of the range

- Number of ideas that are moving from one stage of the process to the next

- Deviation between initial expected financial value of the idea and the ultimate realized value

- Resources expended, both per specific idea and on average

- Resources expended on ideas that move through the process to a particular point but then are not ultimately commercialized

- Performance by the organization with regard to the specification of the process, such as attendance at key meetings, percentage of documentation submitted on deadlines, and percentage of projects passed on to the next stage in the process without having met all of the specified requirements

Cash payback. Most important, management needs to determine whether the innovation process is generating payback. Many companies, even well-respected ones, are not able to determine payback and express it in a way that can easily be understood. The cash curve can be used to "make the numbers come alive" so they can be more effectively managed rather than just reported. Effective measures need to be both clear and ones that management is willing to hold people accountable for. As one senior executive put it, "I have to find one billion dollars of new growth each year. If I don't, then they won't need me the next year."

Indirect benefits. The indirect benefits do not lend themselves as easily to quantitative analysis as does cash, but they can be evaluated.

For knowledge, companies can keep track of the number of books and papers published by their people, the number and type of conferences they have sponsored and attended, and the number of citations that appear in other works.

To evaluate brand strength, ecosystem relationships, and the organizational impact, companies use survey instruments of various types, third-party rankings, or cross-company benchmarkings. Although these do not produce an absolute value they show trends when administered over time.

Highly innovative companies believe innovation and its results can be managed, and they invest significant time and attention to developing their own, distinctive innovation metrics.

Claus Weyrich, member of the managing board of Siemens AG, head of corporate technology, said, "I'd like to contradict the belief that 'if you try to measure everything and do it systematically, it kills any creativity in people.' I do want to have measurements and milestone trend analysis for each project. But of course I know that if it's a risky project, sometimes it might fail. We might run out of time or run out of budget or we do not reach the goal. For me this is not a problem because I know that I have to take certain risks. What's important for me is that people commit themselves to reach certain goals within a certain time. The real innovators don't have difficulty doing that. Of course they would have difficulty if they pass the time frame by one week and I kill the project. This could be counterproductive. But if they do understand that I'm asking for them to commit themselves for certain goals, they are quite happy with it. And they know that if it runs out of time or runs out of budget, I'm not going to blame people, but rather find a way to help them. And this is, I think, the motivating part of measurement."[32]

Mr. Doh-Seok Choi, president and CFO at Samsung, understands very well the connection between measurement, innovation, and payback. "I would underscore the fact that, whatever the process, including innovation, one should set up a rule to be followed and systemize whatever you are doing," he said. "All the managers in the company, in our subsidiaries, as well as in the companies which are cooperating with this company, have a unified system in place. We track performance carefully, the benchmark information is all standardized, and we ensure we meet our targets."[33]

Philips, the giant consumer electronics company, has a remarkably refined approach to measurement, replete with organizational distinctions and metrics of its own devising. Ad Huijser, executive vice president and chief technical officer at Phillips, said, "In 2001, we came to the conclusion that we had spent too much time measuring our mature businesses and not enough time measuring our newer ones. So we started to track, per division, the R&D efforts in three categories 'emerging,' 'growth,' and 'mature,' depending on the market characteristics of the products these were aimed at. As we expected, initial data showed that business managers are inclined, in order to defend their existing profits, to assign relatively more R&D resources to mature business than to emerging and growth areas. So, for each division, we now measure new product sales in relation to total product sales but we don't compare one division to another because the market dynamics of each division are different. For measuring purposes, product life cycles in our Consumer Electronics division are set at one year, but in our Lighting division at three years. Five years from now, that could easily change to half a year and one year, respectively."[34]

To measure the success of its new products and services, Philips has deployed throughout its divisions a standardized equation that produces an "innovation index" number (see figure 7-1). The index is focused on one thing—the payback of the new product or service. The innovation index should exceed one. If it does not, the product is likely a cash trap.

In addition to the innovation index, Philips has a number of other measures, including metrics for product development efficiency, portfolio effectiveness, and the status of key projects. They measure all the critical aspects of innovation performance, from ideation to payback.

FIGURE 7-1

Philips innovation index

$$\left[\frac{\text{New product sales}}{\text{Total sales}} \right] \times \left[\frac{\text{IFO* + R\&D spend}}{\text{R\&D spend}} \right] = \text{innovation index}$$

*Income from operations.

BMW: An Almost Perfect Alignment

BMW has created a remarkably aligned organization that supports innovation and financial results at every level, from strategy to measurement.

"We have a three-level innovation management process here at BMW," Martin Ertl, leader of Innovation Impulses at BMW, told us, explaining, "Innovation is being managed from a central department, and we have strategy satellites in all of the specialist departments, which we call KIFA. *K* is for body shell. *I* is for the integration. The *F* is for chassis, and the *A* is for the drive train. We then have seven innovation councils, each one formed from members of all those KIFA specialist departments. Each one looks like a classical simultaneous engineering group, but also includes people from the marketing department, from the sales department, from the financial department, from human resources, and, of course, from production. But those innovation councils are special because they are not set up with the focus of a certain specialist department, such as engine development or the body shell, but rather to functions that are 'in the eye of the customer.' Each council is led by a manager from the top level of the company. The strategy departments of the BMW group create a strategic framework and this framework gives us our focus areas for the year, our guidelines for the annual process. Before we kick off the annual process, this strategic framework has to be nailed down. Then we have the annual cycle, where each council consolidates ideas and creates a top ten list. And then there is one big day when all the top ten lists are put next to each other and we have to come up with one final top ten list. The problem is, there are always a lot of nice, nifty ideas!"[35]

The councils approve a certain number of projects, and then fund them. To develop a business plan for the proposed project, the managers gather as much information as they can from members of the councils. "You ask the guy from sales, for example, what kind of take rate do you estimate for that and what kind of overall volume do you see?" Ertl said. "And what kind of price do you see for that? And step by step, you gain commitment from people. That is why we work from the very beginning in this multidisciplinary team. We've got to make sure we believe it will generate a return."

Another important element that feeds into BMW's innovation process is the so-called impulse project. Ertl explained, "The main purpose of the impulse project is to bring innovation into the company, especially to look at areas that are not familiar to us. We want to provoke people to do projects that are out of line with what we normally do, that may astonish people, but probably excite them. But they will also sometimes fear them. And that is also something we want to do— look for partnerships and huge potential opportunities. Within automotive, for example, there are always questions about what kind of gadgets might be integrated into a car. But we also want to think beyond BMW as an automotive company. In fact, mobility is a main focus for BMW. And mobility, in addition to automobiles, could also be a boat or an airplane."

The impulse projects have a long history of provoking change and innovation within BMW. Said Ertl, "In the early 1980s, we had a design model for a 3-series station wagon. There had never been a BMW station wagon. The marketing people, and some of the board members, said, 'We don't believe that a station wagon is the right niche for BMW. It's too much of a family car. We're not going to do it.' But then the head of the body shell department said, 'Well, I believe in this idea.' So he collected many parts from existing car models, a rear hood from a Volkswagen Rabbit, and others, and put together a body shell that was more of a touring car than a family station wagon. He took this put-together concept and showed the board and they liked it. They said, 'Let's give it a try.' Today, our touring models account for a big number in our sales total. So there are some successful stories that come out of those impulse projects. But they can also be freaky. Sometimes they don't lead to a lot of sales. But we do them just to show that BMW engineers here are capable of thinking over the fence and are not always stuck in their day-to-day business of automotive, automotive, automotive. If you start giving people the freedom to run and to develop their ideas, it can get very interesting."

As the examples of Citigroup and Samsung demonstrate, alignment is critical to achieving payback. However, alignment does not just happen. Nor is it an activity that can be completely delegated to the HR people or managed solely by an innovation facilitator or innovation unit. Gaining alignment requires leadership.

EIGHT

Leading

If you can decide by the numbers, you don't need a leader.
—Peter Ottenbruch, member of the management board, ZF Sachs

INNOVATION is not a black art, a roll of the dice, or a creative free-for-all. Innovation is an act, not an idea. It requires continual change. It is a combination of deliberate risk taking, close management of a well-defined process, alignment of all the elements of the organization, and an unrelenting and highly disciplined focus on achieving payback. It is a place where determined leaders can make a difference.

Leaders can do certain things, and decide certain things, that others in organizations cannot. Effective leaders realize this and focus their time and attention, their most precious resources, on the things they must do and the decisions they must make.

H. Lawrence (Larry) Culp is CEO of Danaher Corporation, a company that creates a wide variety of instrumentation and advanced industrial products. Danaher comprises many business units, companies, and brands and has achieved remarkable growth, profitability, and share performance since its founding twenty-two years ago. The founders' original intent was to take Japanese manufacturing philosophies and apply them to U.S. manufacturing businesses. The current CEO has taken a similar approach to innovation. Today, many of Danaher's brands are leaders in their industries.

Larry Culp joined Danaher just after receiving his MBA from Harvard Business School and quickly rose through the ranks, running several of the company's businesses—including Fluke, the best known of them—before being named CEO in May 2001, at the age of thirty-eight. When Culp took the helm, Danaher had long been known as an expert in manufacturing. The Danaher Business System (DBS) gave everyone in the company the tools to improve the performance of the manufacturing processes. However, while Danaher became extremely adept at acquiring companies and improving their manufacturing operations, the company recognized that its portfolio of businesses would require an innovation skill set in order to continue to thrive and evolve.

Culp recognized that Danaher was missing the opportunity to create even more shareholder value by increasing organic growth. He believed that he could leverage the manufacturing capabilities, brands, channels, and customer relationships of the various businesses and focus them on developing new products and services.

But Culp was initially concerned that the company's highly disciplined approach would be incompatible with all the qualities that we typically associate with innovation—such as creativity, openness, and collaboration. He realized that he would need to create an innovation process at Danaher that would be "disciplined" rather than "loose" and that would be focused on payback. "One of our engineers put it well," Culp told us, "when he said, 'We don't want to get lovey-dovey about innovation.' On the contrary, we want to make sure that we have clear objectives and hard targets and that we're delivering against them. We need to be thinking clearly about the financial math, as well as the strategy, and not just falling in love with an engineer's great idea."[1]

Given the cultural evolution required, Culp realized that delegating the task to the chief innovation officer would not work at Danaher. "Innovation is everyone's job at this company," he told us. So he decided that he would personally lead the effort. Four years into the initiative, Danaher is a far more innovative company than ever before, with results to prove it. During Culp's tenure as CEO, Danaher's revenues have increased from $3.8 billion to $10 billion. Over the last two years, Danaher's growth platforms have delivered organic growth above 7 percent, which is approximately 33 percent greater than the company's historical rate. Profitability is up 318 percent, and share price has doubled from the low $30s to the low $60s.

True to Danaher's disciplined approach, Culp is very clear about the practices, attitudes, and tools that have enabled the company to embrace innovation and succeed at it. To make innovation easier for the business units, the DBS office developed a set of growth tools to help them better understand customer needs and identify innovation opportunities. To ensure that the budgeting process did not get in the way of innovation, Culp established a "breakthrough funding" process. If a business unit has an idea that it believes can create at least $30 million in revenue (with appropriate payback) over a three- to five-year period, its leaders can propose the idea at the monthly meeting of the Office of the Chief Executive (OCE). "We carve out time at every OCE meeting to review these efforts," Culp said. "So when a president comes in with a good innovation to be funded and says he or she can't find the funds needed to pursue it, if we think it's a winner, we'll give them some help. We have approved forty such projects."

Danaher also established new metrics, required regular reporting, and makes people accountable for their innovation work, but recognizes that performance evaluation needs to be different than it would be for general business activity. "If Barb Hulit, who is the president of Fluke, comes to us and says she wants to consolidate a factory and it will save a million dollars, we'll hold her to that," Culp said. "But if she comes to us and says she wants to spend the same amount of money on an organic growth opportunity, we recognize her sales forecast may have more variability, more risk. If she signs up for $100 million and we hit $85 million, we recognize that it still might be a win in the scheme of things. Or $100 million may not be enough if the market takes off. We hold Barb accountable for the discipline around execution when the decision is made, and judge the $100 million in context. Our demand for accountability is just as strong, but our evaluation is somewhat looser, more judgmental."

Culp has also created a kitchen cabinet of informal innovation advisers. "I cut through the layers between me and the better scientists and engineers in our businesses," he explained. "I'm not talking about just the senior people, but also the people that are literally in the lab every day. Having a conversation with them on a regular basis has been very helpful. Plus, we host an annual innovation conference where we bring one hundred of those folks together so we can talk about what we are doing, what's working and what's not, and network with each other. We've created a community here and it's one piece of the puzzle that works for us."

Larry Culp's experience at Danaher shows that an innovation effort has to be managed in ways that fit the company—its people, businesses, and processes. As he put it, "We think you need to use the culture and the DNA of the organization to drive the innovation process."[2] As a result, every innovation journey will look and feel different from every other. But through our experience and our research, we've found that in all innovative companies there are a small number of areas where leaders can have a fundamental impact on the success, or failure, of an organization's activities and its ability to achieve payback, including:

- Convincing an organization that innovation matters

- Allocating resources

- Choosing the innovation business model

- Focusing on the right things

- Reshaping dynasties

- Assigning the right people to the right place

- Encouraging and modeling risk taking

Because only the leader can decide and act on these issues, they will shape the results of your time in the position—the impact you have, the strength of the organization you leave behind, and your personal legacy.

Convincing an Organization That Innovation Matters

Many organizations, and the leaders who guide them, find themselves facing a challenge that has arisen from the focus on cost, quality, productivity, and effectiveness that has characterized the world of business in the past decade. Although it is now back in vogue, organic growth, and the innovation that drives it, has long been out of favor in many companies and industries. As a result, many marketing, research, and other "nonessential" departments have been gutted because they did not directly and immediately contribute to revenue. In addition, companies have allotted fewer resources to creating the conditions that catalyze idea generation within the company—such as time and space, and the other conditions we discussed in chapter 7—or that enable companies to identify potentially interesting "outside" ideas and build connections and relationships to them.

When a leader begins to talk about the importance of innovation within a context that does not facilitate and reward innovation, people within the organization are understandably skeptical or, at least, need to see a fair amount of proof before they will really listen and even more before they are willing to act. They may not actually be hostile to the idea of innovation (as they might well be to another round of cost cutting or another efficiency initiative), but they have learned from experience to be wary of programs that are described in one way ("A great opportunity for us all!") and executed in another (more work for fewer people). As one employee in a major company said, "It's not that we don't want to believe, it's just that no one has ever made a career here by supporting innovation. But a lot of careers have been broken by doing so."

In our 2006 BCG/*BusinessWeek* "Senior Management Survey on Innovation," we saw a consistent difference between the responses from CEOs/presidents and people at other levels of their organizations. For example, 81 percent of CEOs see innovation as one of the three most important priorities for their company; 61 percent of people below that level agree. And while 76 percent of CEOs believe their companies foster organizations that promote innovation, only 49 percent of employees agree with this view.

Employees need to be convinced that top management is genuinely behind innovation, not only because innovation is a different business approach than they're used to, but also because they know that innovation requires a particular managerial touch. Innovation— and the inevitable mistakes and "dead ends" that go with it—will result in dry periods and uneven progress, and the leaders of the effort must have a strong conviction in the rightness of the path if they are to stay the course during times of uncertainty. Employees will have a very hard time believing that an executive formerly known as the King of Cost Cutting has suddenly become a True Believer in innovation.

Steve Luczo, chairman and former CEO of Seagate, knows this from experience. "I remember someone telling me it would take five years to change the organization," Luczo recounted. "And I remember laughing at him and thinking, we'll do it in six months. But he was right. It took five years." When Luczo took over, Seagate had significant organizational issues that it had to address. "It wasn't, for most people, a fun place to work. You felt, where is the payoff? People really worked hard, and Seagate was typically the leader. But it was tough, because there wasn't any obvious payoff in terms of consistent earnings."[3]

Seagate had, however, an extremely talented and dedicated workforce, Luczo noted, and one fairly adept at reacting to change. So his management team set about restructuring and aligning the entire organization. Most important, the key groups (head, disk, and drive) reported to the same person, so their innovations weren't focused on a particular area but on making a better product Seagate could sell in the market.

At the same time, Seagate's managers and employees had to be convinced that worrying a little less about cutting costs and a lot more about innovation wouldn't come back to bite them. "Getting people to believe in what they did on a daily basis was a challenge," Luczo told us. "Because you were asking people to make decisions that maybe meant things were going to cost a little bit more today, but it was the right decision for flexibility or profitability later. And in the old days, people would get fired for that. There was a bunch of negative reinforcement in place that prevented people from really believing it. But what they saw was that people weren't getting fired—in fact, it was the opposite. People were getting fired if they weren't on the program to get the company structured differently and focused more on innovation."

Communications

When you are a leader driving the innovation process and working to achieve payback, everything you say about innovation matters. You need to communicate your view and commitment in a convincing, compelling, and consistent manner. One CEO we know believes that it takes three years to successfully communicate any new message throughout an organization. In the first year, the CEO and the leadership team hear, understand, and internalize the message. In the second year, members of the next layer down, along with members of the rank and file, get the message. Finally, in the third year, middle management comes to understand and believe what has been said in the previous two years. This CEO also said that by the end of the second year, he was bored by the effort of repeating, over and over, the same message to different audiences. But when results began to show in the third year, he realized again how necessary and worthwhile the continuous communication efforts are.

Holger Schmidt, president of Siemens Medical Solutions, Special Systems Division, echoed the point. "Communication is key, and it starts with me. We say innovation is our key to differentiate ourselves in the market, so people hear that message everywhere. If you went and talked to the accounting department about what makes us different from our competitors, they would all say the same thing: innovation. Even the accounts department understands they need to find ways to contribute to innovation."[4]

Mr. Doh-Seok Choi, Samsung Electronics' president and CFO, told us that changing people's perceptions within Samsung has been one of his biggest challenges. "The biggest impediment has been the existing perception of people. One way to overcome the hurdle is to unify the language and to share the same thoughts. Obviously, that is not going to be easy. I've been working to unify the thoughts and the language for the past thirteen years. I would say I'm about 70 percent of the way there," Choi added. "But I believe that innovation can only succeed from the top."[5]

Daily Activities

Delivering the message about the importance of innovation is only one part of convincing people. "A leader must truly walk the talk," said Claus Weyrich, member of the managing board of Siemens AG, head of corporate technology. "If you talk about innovation, you have to give people the impression that you mean that this is very important for you, personally. That you are personally getting excited by innovation."[6]

Your commitment must be expressed not only in your communications but also in the activities you undertake and the way you allocate your time. If your employees were to look at your calendar, what would it say about your priorities? How much of your month is devoted to innovation activities? How much of the agenda of each event and activity is related to innovation?

Even executives who are earnestly committed to the idea of innovation are often surprised when they do a simple tallying of the actual hours they spend on it. Leadership is always a constant struggle to prevent the urgent from crowding out the important, so innovation is likely to get less attention than it should unless you have truly and deeply committed to it as a priority. Employees are exquisitely attuned to what management does (both when the actions conform to, and when they deviate from, what they say), and the clearest way to indicate the importance of innovation is by spending time on activities that support innovation and engaging in discussions about it.

Whirlpool has worked hard, over a period of years, to align the entire organization around its strategy of innovation. In the words of one senior manager, "From the top of the company, the continued talk and focus on innovation seeps down, and it changes everybody's thinking a little bit at a time. The 'drip' process of an innovative, customer-focused mentality has succeeded because the leadership has never backed off. Every day, there's something about innovation on our portal. In our quarterly meetings, we make it a point to talk about innovation, through our own conversations and by inviting outside speakers who are experts in various aspects of innovation. People knew they were expected to attend transformation meetings where we talked about innovation and how we could do things differently."[7]

Allocating Resources

One of the most difficult tasks for the leader is allocating resources, especially when it involves a change in who or what receives the allocation, or when the allocation is for a project that has a high degree of risk. Most items in the corporate budget are largely fixed and do not change dramatically from year to year. The leader merely reviews these allocations and asks for modifications, if necessary.

But when a major investment in a new product or service is contemplated or a substantial shift in resource deployment is required to increase innovation overall, the task of allocation gets more complex.

The most difficult allocation challenge comes when the company contemplates making an investment in knowledge.

If the knowledge required is product specific—that is, knowledge that will be applied directly to a product or service that is already more-or-less defined—it is likely that the investment is relatively small and the payback reasonably certain, and the allocation decision, therefore, will be easy and relatively uncontroversial. When the company contemplates the acquisition of greenfield knowledge, however, the decision gets difficult. There can be no certainty that the knowledge will lead to a commercial offering or when that might happen, if ever.

"I believe that research is not so expensive," said Didier Roux, vice president of R&D at Saint-Gobain. "When you think about ways you spend money in a company, you have much better ways to waste your money than to do research, even unsuccessful research. You have to accept when you do research that you don't get 100 percent of success in a short period of time. It's an advancement the same way as when you do marketing, or you have ads or commercials on the TV; sometimes you don't get 100 percent of the success you expect. Nevertheless, you have some impact, and you correct for the next campaign, and it is more successful because you have already done one. You have to have some percentage of nonidentified or not-well-identified money; otherwise you take no risk, and if you have no risk, you have very little chance to get a breakthrough."[8]

Resource allocation for gaining greenfield knowledge is a fundamental issue with long-term implications for the success of the company as a whole. Deciding which areas to invest in and how much to invest in them, in fact, goes to the heart of the strategy of the company. And there will be many differing viewpoints on what to invest in, when, and how much to allocate. Where do you think the market is heading? In what areas do you want to compete? How are you going to compete and win there?

This can be the most difficult of the decisions a leader must make about innovation, because it requires developing a vision of where and how the company will compete in the future, and determining what new knowledge will be needed to do so. In other words, the leader must place bets—well-informed, well-reasoned, deeply felt bets—but bets nevertheless. And the leader must be willing to stick with these bets long enough for them to have a chance of paying off, but not so long that they become a dangerous drain.

Bill Gates has placed many such bets. "At Microsoft, we have the extreme luxury of being run by a chairman who has learned a couple lessons," Bill Mitchell of Microsoft told us. "One of them is, 'Sometimes it takes two or three versions to hit that home run.' He's got patience and the long view. And the other lesson is that something that starts really small can sometimes blossom into something really big. And it's difficult to predict which ones those will be. Because he's had those life lessons multiple times, he is a great sponsor."[9]

Choosing the Innovation Business Model

The choice of an innovation business model is also intimately connected to the company's choice of a strategy, and the company's leaders must be involved in the decision. The choice of model not only is a matter of strategy, it has a significant impact on the probability of achieving payback and the timing of that success.

"Sometimes we can't find a good business model for a particular new technology in our company," Claus Weyrich, at Siemens, told us. "So we try to enter a joint venture with somebody else, or we form a company, try to get venture capital, or we sell and license the technology to somebody who then gives us a certain strategic advantage by selling those products to our company. There are many ways to go."[10]

The leaders must ensure that the company and the management team consider all the options. Too often, a company simply continues using the innovation model it has always used, even when the model has lost its ability to innovate or create payback.

For example, Kim Winser, CEO of Pringle of Scotland, a clothing company, made a bold change of model that enabled the company to reverse its declining performance, achieve payback, and once again become an innovative force in its industry.

Founded in 1815, Pringle was once renowned for its innovations in cashmere knits, the twin-set sweater, and the familiar Argyll pattern. But by the mid-1990s, the brand had slipped to a point close to irrelevance. Having pursued an ill-conceived licensing strategy, the company no longer manufactured most of the products that bore its name and was selling virtually everything at a discount. Remarkably, not a single cashmere item was available with a Pringle label. In fact, all the cashmere cloth produced by Pringle's venerable factory, which employed one hundred skilled craft workers, was sold to a cloth jobber.

In 2000, Winser became Pringle's chief executive, and she made the decision to pursue innovation as a path to organic growth. She decided that the best way to support that strategy was to use the integrator approach, leveraging the skills and capabilities that had originally made Pringle successful. In particular, Pringle owned a number of knitting machines that had long been mothballed but had a functionality few others could match—the ability to create the interlocking patterns known as intarsia. "I flew to Scotland and found all these wonderful machines under covers, with three inches of dust on top," she recounted to us.[11] In addition, many of the workers who knew how to run the machines were still with the company, although working at other jobs. The plant manager asked his father, who had been the manager before him, to come back to help restart the machines, Winser said.

So, contrary to the expectations of many observers, Pringle did not sell off its assets and close its one remaining plant. Rather, Winser revitalized her team of designers and marketers, renegotiated contracts with cloth jobbers, and announced that Pringle would create innovative, high-end products for the luxury market. The company would do so by selling directly to retailers and opening its own signature stores. Winser began terminating licenses and ending relationships with over one hundred accounts that she saw as brand damaging. Once, Winser said, while on a trip to Japan, she stumbled across an advertisement for a Pringle-licensed fly-fishing rod. "I didn't even know that license was out there," she said. "We ended it as soon as we could."

The strategy worked. The Pringle brand was soon revitalized and seen as an important player in the fashion market. Winser's move to the integrator model, along with the other steps taken, has enabled Pringle to reposition the brand, significantly improve cash payback, return to steady growth, and had dramatic organizational benefits. "We redeveloped the business, the factory, and cashmere," Winser said. "The licensing had been done out of desperation. Now, we're opening our own stores all over the world."

Of course, the strategy was not without risk. Winser had to jettison the source of much of the company's then-current revenue, change strategy, rebuild a set of innovation capabilities that had been allowed to atrophy, and bet against the prevailing wisdom in the world of apparel, which overwhelmingly follows the orchestration model, especially for companies in high-cost locations such as the United Kingdom. Only a leader in her position could have taken such necessary, and successful, actions, and it has paid off.

Pringle was making substantial losses in 2000. Now, although exact profitability is a closely guarded secret, Winser allows that revenue is ten times what it was just a few years ago. What's more, the brand has regained enough luster that Pringle has been identified as a potential acquisition target by other well-known luxury companies.

Focusing on the Right Things

As organizations grow larger and more complex and they consider a variety of avenues for new sources of growth, the number of ideas, projects, proposals, and innovation units proliferates. Leaders of highly innovative organizations must spend a great deal of time determining which ideas will be pursued or left alone and how those decisions will be made. They must focus the organization on thinking about, developing, and commercializing the right things. They must stop things that are not going to generate payback. The worst possible outcomes will result if these decisions are not made by the leaders, but are left to be made by others.

In our work, we see that senior management wants to be involved in focusing an organization on the innovations that will have the greatest impact, and they try their best to be involved in the process. But they often are sufficiently distant from the details so that when they do weigh in with their opinion, it comes across as either uninformed or biased. As a result, projects often get selected and advanced not because they are necessarily the best or right things to do, but because "senior management wanted them to move through."

Strong innovation leaders don't just "move things through." They stay close to the action and build consensus around new ideas and projects. Mike Snyder, former president at ADT Security Services, the world's leading maker of security systems, told us about Les Brualdi, the former CEO of ADT and the godfather of the most important innovation in the home-security business—the mass-market residential security system. "Les always led from the front. He never wavered. Even when we'd go to market with this crazy new offering and nothing would happen right away, he never took his hand off the tiller. But he always enlisted consensus and support, too. He would bring groups together on a frequent basis to review results and best practices. They would always come up with one or two important tweaks. He was a master at not listening to and not supporting the stuff he didn't agree with, and reinforcing and emphasizing the stuff that he wanted to move forward with. But he could do it so that the group accepted it. That's what a leader does."[12]

Not only does "focusing on the right things" mean starting up the right new initiatives, it must inevitably involve putting an end to those things that are not going to generate payback. The decisions about which innovation projects to continue and which ones to deprioritize are always important, emotional, and difficult. Without strong leadership, most organizations can't make these decisions effectively. Instead of pruning away the deadwood so they can focus on the projects that have potential for success, they endlessly try to fix up the ailing projects, rehash the problems and possible solutions, and second-guess what went wrong. As a result, they lose the ability to reach closure and mass support behind projects that will result in cash and indirect benefits.

"Balancing passion and objectivity is my biggest challenge," said Andreas Gutsch of Degussa. "On the one hand, you need to be passionate to be innovative. But on the other hand, we need to stick to our plan, and if we see our plan is not materializing, we can try to fix all the problems and do our best to make it a success. But if we fail and we see the project is running up against the wall and is likely to crash, we need to stop it. And this is where we get exhausted. It is a very tough situation when you've given everything you can as a team to win; but you have to stop it because that's the rational decision. The emotions of the people involved, and even those just watching, are very affected by the decision. It is very difficult and you have to be prepared to argue very hard, even to your own board, to stop the project—it's not so easy."[13]

Reshaping Dynasties

Only a leader can assess the role that a dynasty is playing within the organization and reshape it when necessary.

There are two problems with dynasties. The first is that the dynasty tends to suck up as many resources as it possibly can and may therefore starve innovation initiatives that exist outside the dynasty. The second is that as dynasties grow, they tend to become less committed to innovation and will often seek to eliminate the new. There also tends to be a significant lag between the time when a dynasty's payback begins to lessen and when its allocation of resources lessens proportionally.

Reducing the influence of a dynasty may require cannibalizing its successful offering with the introduction of a new one that directly competes. This will always be opposed by a dynasty. When threatened, it will do everything it can—things both subtle and not so subtle—to put an end to the threat. It may not be necessary, however, to put resources into a cannibalistic new offering to reshape a dynasty. Simply boosting the allocation to a noncompeting new development area may also shift attention, influence, and power away from the dynasty.

It is the leader who must make the decision to reallocate resources, both financial and, most important, human talent, away from the dynasty toward young, potentially threatening, and promising new areas. To do so, the leader must learn how to respond to the ways that people within the dynasty will defend themselves and it. They may argue, for example, that any change in resource allocation to the dynasty will have a negative impact on the entire culture, because "this is the product that built the company." They may get more personal and suggest that "careers are made" in the business of the dynasty and, by implication, can be scuttled by actions that harm the dynasty.

Teruaki Aoki, senior executive vice president and executive officer at Sony, helped challenge the dynasty of cathode-ray tube (CRT) color televisions at his company. Introduced in the late 1960s, Sony's CRT sets generated tremendous payback and helped Sony become the largest manufacturer of color TVs in the world. "But we knew that LCD and plasma were coming," said Aoki. "However, in the mid-nineties, they were not yet real threats, and our CRT business was still extremely successful. That success delayed our shift to LCDs and plasma."[14] It proved extremely difficult for Sony to deemphasize a business that had been so successful for so long.

"CRTs had been a great business for Sony," said Makoto Kogure, the senior vice president who heads Sony's TV group. "But then we had to make investments elsewhere."[15] Starting in 2004, Kogure gradually shifted resources away from the CRT group and refocused the organization on new and emerging technologies. When we visited Sony's Global Television headquarters, we were shown into a special room packed with some of the most advanced television units in the world—

both production sets and prototypes—and not a single one was based on CRT technology.

Generally, the leader is the only one who can reshape a dynasty that has lost its ability to innovate or no longer can generate the required payback from the resources it is allocated.

Putting the Right People in the Right Place

Ultimately, innovation is about the people involved in the process. Successful innovation requires the internal orchestration of many different functions, disciplines, geographies, and activities. People either play a positive role and help organizations develop and turn great ideas into cash payback and noncash benefits, or they get in the way of doing so. Rarely is the effect of an individual benign, at least in terms of the outcome. The right person in the right place with the right skill set and motivation and approach can make all the difference. The converse is also true.

Over the years, many of our clients have commented that there are "horses for courses," meaning that different people have different strengths and vary in their abilities to manage different situations. When leaders undertake the task of improving the performance of their innovation activities, the effort frequently involves the identification and closing of capabilities gaps—both in their organizations and in themselves. Effective leaders of innovation (and, in fact, anyone involved in the process) tend to possess a set of qualities and skills that may be less necessary for leaders of other types of efforts. These qualities include:

- **Tolerance for ambiguity.** The innovation process is inherently ambiguous. It is filled with uncertain outcomes, complex relationships, multiple possibilities, and conflicting ideas. People who prefer clear-cut directions, predictable outcomes, quantitative measures, and well-defined relationships will not be the best leaders or managers of innovation. As Renault's Yves Dubreil, director of R&D, put it, "Innovation will always have uncertainty and ambiguity. You must be able to deal with ambiguity. It is very closely related to risk, and the ability to accept uncertainty is critical to leading innovation. If you want to be certain, you do anything else, but not innovation."[16]

- **The ability to assess and be comfortable with risk.** Risk is always part of innovation, and leaders must be able to tolerate risk themselves, as well as encourage others to take risks and help them learn how to do so. A leader who can tolerate the risk involved in innovation is usually one who feels confident in himself and is sure of his professional abilities. He must be steady enough to withstand a few hits during the process of innovation; be able to tolerate the questioning, debating, and second-guessing he is likely to encounter along the way; and secure enough to put his own short-term career considerations second to what's best for the company and the new product or service he is helping to create. Risk taking is decidedly not for the faint of heart.

- **The ability to quickly and effectively assess an individual.** In betting on specific innovations, one is often betting not only on the underlying strengths of the idea but also on the leader who will drive the effort forward. "We have offices where the innovation is led by guys who are actually poor managers but excellent leaders," Snyder at ADT told us. "We also have offices where, whenever we try to innovate anything, it never seems to stick because the people are very good managers but poor leaders."[17]

- **The ability to balance passion and objectivity.** No idea has ever been successfully commercialized and realized without somebody being passionate about it. In most organizations, copious amounts of passion are required to overcome organizational inertia, push an idea through the entrenched bureaucracy, reshape a dynasty, and get past the multitude of barriers that will be encountered along the way. While it is often said that there is no dishonor in making a mistake, there is often a high price tag for doing so, especially when the mistake takes the form of a project that is allowed to go on for too long.

- **The ability to change.** Leaders who are successful at innovation are willing to change. Sometimes the required change is on an organizational level. You realize that people, processes, and structures that used to be successful no longer are. They need to be changed, and someone needs to have the insight to see the required change, an understanding of how to make the change happen, the courage to attempt it, and the determination to see it through.

  At other times, the change is on an individual level. Although some leaders are born with a fervent desire to constantly create new things, most come to innovation from some other viewpoint and discipline. Understanding your own attitude toward innovation is critical, because it will affect how you build your organization and how you interact with the process. All leaders that we consider to be exceptional at managing innovation are willing to change as individuals—their attitudes, opinions, activities, routines, and behaviors—and can usually describe with a lot of detail the many times they have done so. Do not expect to end the journey of innovation as the same person you were when you began it.

When the leader determines that people currently in management positions do not possess these innovation-friendly qualities, it may be that the fit is no longer right. One of the most difficult decisions that a leader must make is to move employees out of positions where they once fit well and ask them to move on—either within the company or beyond it.

Only a leader can make sure that the organization has the right people in place to lead it through the innovation process and help it achieve payback. Mr. Young-Soo Song, vice president of human resources at Samsung, put it this way: "A major strength for Samsung is the importance we put on people. Chairman Lee has a very strong mind-set. He ordered every CEO in the company to hire the best people in the world. So we brought in some people who were paid more than their CEOs. And Chairman Lee personally tracks them. So if you have that one person in the organization, he can change everyone in their own area. Hiring the right people is very important, but it is even more important to put the high-potential people in the right place. Chairman Lee believes that, in the twenty-first century, the right person should be able to feed one thousand or even ten thousand others."[18]

Encouraging and Modeling Risk Taking

For many companies, the central responsibility of the leader—and the most difficult part of innovation—is dealing with risk.

Entrepreneurs and small companies may thrive on risk (or seem to, probably because they don't have much to lose), but executives and large companies—especially publicly held ones—generally do not. Companies create structures and processes and metrics for the very purpose of measuring, monitoring, and eliminating risk. Few people within large companies make a successful career by associating themselves with risk. The skills that enable people to ascend into senior management are more associated with risk avoidance than with risk taking.

So when a company sets off on a course of innovation, it faces a fundamental problem: how to make people understand that risk taking must be part of the innovation process and encourage them to make decisions and take actions that accept some risk rather than rejecting all risk.

This is an effort the leader must be involved with. Pattye Moore, former president of Sonic Corporation, one of the most innovative quick-service restaurant chains in the United States, told us, "Somebody in senior management has to be a risk taker herself. If you don't have someone like that, bureaucracy will take over. People will fear risk taking. You better have somebody in your senior management team that loves to innovate."[19]

But the presence of a few risk takers and true believers in the top ranks of management will not automatically ensure that others within the organization will change their views and behaviors relative to risk taking. This is because employees know there are still plenty of other members of management—among them, no doubt, their bosses—who are more focused on achieving short-term goals, particularly quarterly numbers, and will find ways to squelch any risky ideas that do not contribute to those goals, no matter what the chief innovator says or does. Accordingly, employees are afraid that if they take a risk and it fails, they will face unpleasant consequences.

Thus, there is an idea stalemate, with senior executives asking for more new, bigger (and therefore riskier) ideas, and employees wanting to propose them, but both "sides" unable to break out of the risk-averse mode.

Peter Ottenbruch, member of the board of management, Powertrain Division at ZF Sachs AG, said, "I wish our organization took more risks and that the ideas we pursued were bigger. To change this, we tell people stories about times when we didn't take enough risk and were too slow, and we lost opportunity. We try to make sure that while we look at the numbers, we also make decisions based on saying, 'We believe in this.' You have to be willing to go beyond simply making decisions based only on the numbers."[20]

In our experience, the company that learns to assess and live with risk will achieve a significantly higher rate of return than companies that do not. An idea that is chosen deliberately, whose goals are clearly defined, and that is commercialized and realized with rigor, but still does not produce the desired return can rarely be considered a complete failure. "Intelligent failure is the way you react and learn—you learn more from failure than success," said Dubreil of Renault. "But if you want to learn, you have to work at it. You have to spend sufficient time and intelligence in checking why the failure occurred to understand it completely."[21]

Innovation always involves risk. It can be hedged, mitigated, insured against, and shared with others, but—by definition—can never be eliminated from innovation, nor should it be. Leaders and companies that will not take enough risk will never be able to achieve payback through innovation. Companies that say they want to innovate, but then do everything possible to remove all the risk, will confound the process and drive innovative people away.

Companies that accept the risk, and commit to managing the process holistically, will find that innovation not only can be an endlessly renewable source for generating payback, indirect benefits, organic growth, and longevity, but also can create a positive legacy for the company and its leader.

In our experience, the greatest risk that leaders and their companies face is taking no risk at all.

Taking Action

I F YOU ARE COMMITTED to improving your company's payback from innovation, you have already taken some actions to achieve that goal. Hopefully, in reading this book, you have identified others that you want to try. You know that innovation is a process of learning and improving rather than a matter of making a single decision and stepping back to watch its execution happen.

As a result of our work with hundreds of organizations, we know that the toughest step is the first one. So when executives ask us, "Where do we start?" we answer with six actions that can be taken to begin the innovation journey.

Draw a Cash Curve

Draw a cash curve for a current or upcoming innovation project that you are considering. Force yourself to understand what you need to know to draw it as accurately as possible. As you draw, ask yourself the tough questions, including the following:

Start-up costs. Are they too high? Too low? If too high, could we share them with partners by using a different innovation business model? If they look low, have we invested enough to ensure that we can overcome potential technical, executional, or market risk?

Speed to market. Do we need to move faster? What are our competitors doing? What might they do? What are the market conditions? How long will the market life cycle be for the new product or service? Could we move faster with a different innovation business model?

Scale. How much do we need to sell, and how quickly, to ensure payback? If we miss our volume targets, how great is our risk of not achieving payback? If we exceed them, do we have the capacity to quickly increase volume? Are we employing the right innovation business model to manage possible variations in scale?

Support costs. Have we considered all the costs that will go into launching, supporting, and constantly improving the new product or products? What education is needed? Is there any risk that support costs will be considerably higher than planned and the product will become a cash trap?

While a cash curve looks simple to create, it is actually quite complicated because of all the facts, assumptions, and judgments needed to draw it correctly. But the act of drawing a cash curve forces the discipline of asking the right questions and provides a sanity check on the answers. (For example, do we really think we can get our market share to move from its current 4 percent up to 30 percent in eighteen months?)

Once you've drawn the curve, based on the variables you've gathered, ask yourself if you like the shape of it. If not, think about what you could do to reshape and improve it. For example, does it require taking on more risk or less? Could risk be reassigned from one part of the curve to another? How?

Take the time to draw the curve rigorously. You'll gain a greater understanding of innovation and what it takes to be successful. You'll learn how to use the curve to help you make the trade-offs required to increase the payback from your projects before you start spending money (or to improve the impact of the money you are already spending).

Understand What Is in Your Innovation Portfolio

Get your arms around what is in your innovation portfolio. Most companies do not have an accurate view of the complete contents of their portfolio. If you catalog what innovations are being pursued, what resources are being consumed, and what results you are expecting, you will gain a new understanding of what you are actually investing in. Do a sanity check, asking such questions as these: What are the true cash inflows and outflows we expect? (Draw a curve to show these, even if it seems "too early" to do so.) Does what we are being told make sense? Should some projects be dropped? Should funding be reduced or increased? Are our projects moving slowly because we are trying to do too much with too few resources, or should we be moving faster?

Don't just look at the "official" projects, but look at all of the "unofficial" projects, too—the hundreds of efforts that are taking place because people in the organization think they are a good idea and they can cobble together enough resources from existing budgets.

In our experience, the projects under review will fall into three categories. About a third of them will be winners that should be promoted and accelerated. Another third will be a waste of resources and should be stopped even if they are being supported by "other budgets."

These are the "walking dead," and it takes great courage to kill them, but it must be done. The final third will be less easy to evaluate and will need further exploration and discussion to determine whether they should be kept or killed.

Moving on the third that should be stopped, and reallocating resources to accelerate the third that are winners, will immediately increase payback.

Appoint a Leader or Take Leadership Yourself

Without leadership, little progress can be made on improving payback. Your organization needs to know that you are serious about innovation, and to truly understand that, they need to see leadership.

To demonstrate leadership, you need to either appoint a leader of innovation or be the leader yourself. If you appoint a leader, the leader must be empowered to take action, and you must support that person. It doesn't matter what the leader's title is, so long as he has power, influence, credibility, and authority throughout the organization. Without a leader who can see and act across organizational boundaries and who has the resources and authority to influence the process, capturing the full payback from your investments will be impossible.

When thinking about who might be the best choice for innovation leader, consider the qualities required: a tolerance for ambiguity, comfort with risk, the ability to quickly and effectively assess an individual, a balance of passion and objectivity, and the ability to change.

Rethink the Innovation Models That You Typically Use

Most companies use the same innovation business model (integrator, orchestrator, or licensor) for almost all their projects. Review some projects and consider whether a different model would have yielded higher payback and whether that would have freed up resources that could have been used to create more new products and services.

You'll probably find that you would have changed the mix of models. More important, take a look at the projects that you have in the pipeline now. Why are you using the models you are? What are the barriers to using others? What might the use of different models mean for the amount and allocation of cash, indirect benefits, and risk? What does the cash curve look like for each of the three models?

Hunt for the Cash Traps

Every major company we've seen has some cash traps. They can be existing products that are clearly not generating cash (or indirect benefits that justify them) or projects that are under way that are not likely to generate cash returns.

Take a look at the existing portfolio of products and the development pipeline, and identify efforts that are likely to become traps. Pick a few and dive into them to determine their likely returns. Take a very close look at those that are justified by indirect paybacks, and make sure they are really going to deliver or determine whether they are just cash traps. You are likely to find at least a few that should be shut down. When you shut them down, you'll find that you will have freed up enough cash to fund an expansion of your innovation efforts.

Rethink Your Perspective on Risk

Your perspective on risk can make all the difference in successfully managing innovation. A few questions to ask yourself: Are we looking at individual projects rather than thinking through the risk in the context of our portfolio and corporate ability to absorb risk? Are we looking at real probabilities rather than "worst cases"? Are we looking at the risk of not doing the innovation at all?

As you consider how to get started, keep in mind that the road to achieving extraordinary payback from innovation can be a long one.

But even a small step can improve and enhance payback almost immediately. What matters most is not the exact step you take first, but that you take deliberate, explicit, and committed action to improve your payback from innovation—and that you do it now.

ACKNOWLEDGMENTS

T HE DEVELOPMENT, research, writing, publishing, and supporting of *Payback* have required the will, commitment, engagement, creativity, intelligence, resources, hard work, diligence, and collaboration of dozens of people. To all of them we give our heartfelt thanks; the book could not have been accomplished without your contributions.

First and foremost, we owe our thanks to the many senior executives around the world who spent countless hours with us during our research for this book. They are named in the text and notes. We owe them a great debt of gratitude for sharing their years of experience and learning, and their successes and failures, with us and our readers.

We also owe a great debt to the companies and executives we have worked with and learned from over the past twenty-five years. While confidentiality prevents us from naming them, our deepest thanks go to them. We have been privileged to consult with many of the most innovative companies in the world and some of the very best executives. They have pushed us to develop our thinking, approaches, and abilities, and have served as the whetstone against which our ideas were honed. We have never failed to learn from them, and they are why this book and our consulting practice exist.

Payback would not have happened without the support of two BCG CEOs. Carl Stern and Hans-Paul Buerkner have encouraged us to crystallize our thoughts and have supported us throughout the process of getting our ideas down on paper. They have believed in us and pushed us forward.

John Butman, an independent writer who has helped BCG authors create a small shelf of titles, has played an important role in making this book. He worked closely with us for more than two years, first to develop a successful proposal and then to help us crystallize, clarify, shape, and effectively communicate our thoughts and ideas. His touch with the written word is unrivaled. He has provided expert guidance and wise counsel and was tirelessly dedicated to this project.

Before there was a book or even the idea of a book, there was a small band of passionate people who took an interest in the ideas that would first become BCG's "Innovation-to-Cash" practice, then a *Harvard Business Review* article, then *Payback*. All had vital roles in getting us through those "early days." They included Massimo Russo, Sims Huling, David Panzer, Mike Petkewich, Jeff Gell, Brent Pycz, and Christine Barton. Two early pioneers deserve special mention for their contributions over many years to this work. Chris Mark was there at the beginning and served as our right hand until he moved on to a new career. He was a valued thought partner, tireless contributor and capturer of thoughts, and remains a good friend. James Stark continues to make significant contributions to both our intellectual capital and ability to serve clients, and has led our survey and database development efforts. Without Chris and James we would not be where we are today.

As the early research work evolved into the creation of a book, the core team reconstituted itself; its BCG members included—for varying periods of time—Petros Paranikas (who led the team for many months), Kunal Mehra, Liraz Evenor, Shamik Lala, Keith Rabin, Yuliya Kravtsov, Michael Lasota, and Obi Arinze. Roger Moore, along with Petros, has devoted his effort to the development of The BCG Innovation Institute.

We received expert counsel about the messages and marketing of the book from Bill Matassoni, who until recently was the head of global marketing for BCG. K.C. Munuz continues to provide wise advice, constant encouragement, and many useful insights. George Stalk gave us the benefit of his many years of working with clients, writing books, and thinking about innovation. Chris George was very helpful with approvals and marketing, and Eric Gregoire has been tireless in helping with getting our messages out. Gerry Hill has also contributed to our work, especially our annual survey.

We have been fortunate to have the help of a wonderful agent, Todd Shuster of Zachary Shuster Harmsworth, who helped guide us in the ways of the publishing world. And Jacque Murphy and the team at Harvard Business School Press have added greatly to what you are reading.

Our administrative assistants Lisa Butler, Paula Daly, and Ruth Kohn supported us in this exceptional project with the same diligence and skill they provide us in all of our activities at BCG. Marge Branecki and Maureen Kwiatkowski did similarly admirable work in transcribing many of the tape-recorded interviews.

The entire team was supported by a crew of researchers, many of whom are members of the Knowledge Group at BCG, including Vera Ward, Bill Hagedorn, Jill Jackson, Rudy Barajas, Wanda Perkins, and Simone Bruegel.

Thanks also go to the many people who provided us with help, support, knowledge, or information, in ways too various to enumerate here; they include Alexander Wiegand, Jan Friese, Jinkyoung Kim, Rickard Akesson, Fredrik Burling, Christina Coffey, Erik Flinck, Eric Hart, Christine Vollrath, Lauren Whitehurst, Jack Whitt, Jill Corcoran, Jennifer Balaskas, Andy Forstner, Lauren Moore, Bill Turner, Jack Zhu, Sarah Davis, Mel Wolfgang, Arnoud van den Berg, Erol Degim, Johanna Kahn, Ajit Ketkar, Raymond Nomizu, Rajesh Srinivasan, Doug Turek, Yves Morieux, Dan Grossman, Dustin Burke, and Elizabeth Rizza.

Many thanks to those who helped us secure the interviews that add so much detail and color to the book. They include Professor Nei Hei Park, Sebastian Ehrensberger, Andreas Maurer, Josef Rick, Georg Sticher, Hubertus Meinecke, Pascal Cotte, Carsten Kratz, Michael

Fuellemann, Steve Chai, Christoph Schweizer, Antoine Gourevitch, Takashi Mitachi, Udo Jung, Andy Blackburn, Xavier Mosquet, Stepan Breedveld, Christoph Nettesheim, and Rolf Bixner. During our travels in Korea, we benefited from the help of our translator, Lee Kaphyun, and we were ably assisted in our arrangements and logistics by Yuseon Lee and Katrin Heinemann.

In particular, Jim would like to thank his late father, James L. Andrew, who first taught him about business, and his mother, Ann B. Andrew, who both contributed to and sat through many a dinnertime conversation on the subject.

Finally, we would like to thank the many people whom we have learned from and worked with on these topics over the years, and who have helped to shape what BCG's innovation practice, this book, and we have become. They include John Clarkeson, Kermit King, Arindam Bhattacharya, Matt Krentz, Jim Lowry, Petter Eilertsen, Eric Olsen, Tom Lewis, Mark Freedman, Paige Price, Bernd Waltermann, David Michael, Larry Shulman, Wilko Stark, Monish Kumar, Jim Hemerling, Kim Wagner, Simon Goodall, Atsushi Morisawa, Rune Jacobsen, François Dalens, Patrick Ducasse, Michael Silverstein, Collins Qian, Rich Lesser, Sebastian DiGrande, Amyn Merchant, Paresh Vaish, John Garabedian, Brent Beardsley, Joe Manget, Réne Abate, Pete Dawe, Michel Frédeau, Andrew Taylor, Brett Schiedermayer and the team at the BCG Value Sciences Center, Ralf Dreischmeier, Ken Keverian, Pete Lawyer, Alison Sander, David Dean, Thomas Bradtke, Martin Koehler, Patrick Forth, John Wong, Martin Reeves, Anthony Pralle, Philipp Gerbert, Pete Lawyer, Per Hallius, Steve Matthesen, Marty Silverstein, Ralph Heuwing, Kaz Uchida, Jan Koeppen, Paul Gordon, Knut Haanæs, Marin Gjaja, Jeanie Duck, Luc de Brabandere, Lars Fæste, Mark Lubkeman, David Pecaut, Lars Terney, Anna Minto, Tom Lutz, Ron Nicol, Nick Glenning, Kevin Waddell, Bob Victor, J. Puckett, Nicholas Keuper, Vladik Boutenko, Vikram Bhalla, Massimo Busetti, Naoki Shigetake, Dieter Heuskel, Armin Schmiedeberg, Bjørn Matre, Harri Andersson, Charmian Caines, Renaud Amiel, Steve Gunby, Mary

Barlow, Yvan Jansen, Bruno van Lierde, Jim Borsum, Roland Loehner, David Rhodes, Colm Foley, Simon Goodall, Ian Frost, Gerry Hansell, Ramón Baeza, Doug Hohner, Jesús de Juan, and Dave Young, as well as many other BCG colleagues who have provided support, counsel, encouragement, and advice on our *Payback* journey.

Thanks to all the people listed here and to any others we may have inadvertently omitted; you all have helped to create this book and make it a success.

One: Overview

1. "The World's Most Innovative Companies," *BusinessWeek*, April 24, 2006.
2. Teruaki Aoki, interview by author, June 7, 2005.
3. Young-Ha Lee, interview by author, January 11, 2006.
4. Jim Koch, interview by author team, July 20, 2006.

Two: Cash and Cash Traps

1. Robert A. Guth, "Microsoft, with Fat Dividends, Attracts a New Class of Investors," *Wall Street Journal Europe*, January 27, 2005.
2. Quentin Hardy, "Iridium's Orbit," *Wall Street Journal*, June 4, 1998.
3. "Going, Going, Nearly Gone," *Economist*, September 9, 2000.
4. "The Ill-Fated Satellite Venture Has Re-Launched Itself in More Modest Form," *Economist*, July 14, 2001.
5. George Stalk Jr. and Thomas M. Hout, *Competing Against Time: How Time-Based Competition Is Reshaping Global Markets* (New York: Free Press, 1990).
6. Robert A. Guth, "Getting Xbox 360 to Market," *Wall Street Journal*, November 18, 2005.
7. Stefan Rinck, interview by author, January 2006.
8. Don Remboski, interview by author team, September 10, 2005.
9. Pierre-Emmanuel Levy, interview by author, January 17, 2006.
10. Jeffrey Krasner, "To Cut Costs, Biogen Idec to Sell Drug Plant for $408 Million," *Boston Globe*, June 17, 2005.
11. "Biotechnology—North America," *Mergent Industry Reports*, August 1, 2005.
12. Deborah Ball, "As Chocolate Sags, Cadbury Gambles on Piece of Gum," *Wall Street Journal*, January 12, 2006.
13. Bruce D. Henderson, "Cash Traps," *BCG Perspectives*, 1972.
14. Jim O'Connor, interview by author, May 12, 2005.

15. "History of the Supersonic Airliner," CNN.com, July 5, 2001, http://archives.cnn.com/2001/WORLD/europe/04/18/concord.history/.

16. "Concorde Anglo French Supersonic Passenger Jet Aircraft Development History," http://www.solarnavigator.net/aviation_and_space_travel/concorde.htm.

17. Gene Munster and Michael Olson, "Company Note" on Apple Computer, Piper Jaffray, January 15, 2006.

18. Brent Schlender, "Apple's 21st Century Walkman," *Fortune*, November 12, 2001.

19. Erik Sherman, "Inside the Apple iPod Design Triumph," *Electronics Design Chain*, Summer 2002.

20. TNS Media Intelligence, Ad$pender report on Apple Computer 2006, http://www.tns-mi.com.

21. Peter Burrows, Ronald Grover, and Tim Lowry, "Show Time! Just as the Mac Revolutionized Computing, Apple Is Changing the World of Online Music," *Business-Week*, February 2, 2004.

Three: The Indirect Benefits of Innovation

1. Didier Roux, interview by author, January 17, 2006.

2. Mario Tokoro, interview by author, June 7, 2005.

3. "Inside the Deal That Made Bill Gates $350M," *Fortune*, July 21, 1986.

4. Sheigh Crabtree, "Pixar Suit Claims Patent Breach," *Hollywood Reporter*, March 12, 2002.

5. Pixar Animation Studios, "Pixar Animation Studios and Exluna Settle Lawsuit," news release, July 22, 2002.

6. Frank Rose, "The Seoul of a New Machine," *Wired*, May 6, 2005.

7. Lim Sun Hong, interview by author, January 12, 2006.

8. Wojtek Dabrowski, "Korea Eyes TV-War Victory," *Financial Post*, November 8, 2003.

9. Yun-Hee Kim, "Weak DRAM, LCD Prices Likely Hurt Samsung Elec's 1Q Net," *Dow Jones International News*, April 12, 2005.

10. Stefan Rinck, interview by author, January 17, 2006.

11. "Debut of Kenmore PRO Line," April 21, 2006, http://www.newswire1.net/NW2006/A_WEB_MO/WEBken/.

12. "Singin' the Blus—Standards Wars," *Economist*, November 5, 2005.

13. Ake Wennberg, interview by author, March 21, 2006.

14. Peter Ottenbruch, interview by author, January 21, 2006.

Four: The Integrator

1. Martin Ertl, interview by author, January 19, 2006.

2. "Ideal Employer Rankings—Top 50: Engineering & Science," *Universum Graduate Survey*, 2005.

3. "Intel Manufacturing Frequently Asked Questions," http://www.intel.com/press room/kits/manufacturing/manufacturing_qa.htm#1.

4. ECCO, "ECCO Sko A/S Has Opened Its First Factory in China," news release, April 18, 2005; and "Footwear/Ecco Seeks Wider Market: B200m Step in a New Direction, *Bangkok Post*, November 7, 2005.

5. Adrian Baschnonga, "Nokia Anticipates Strong Mobile Growth," *Global Insight Daily Analysis*, December 2, 2005.

6. Roger Cheng, "Nokia Says Co. Holds 20% Cost Advantage over Rivals," *Dow Jones News Service*, December 1, 2005.

7. Kasra Ferdows, Michael A. Lewis, and Jose A. D. Machuca, "Rapid-Fire Fulfillment," *Harvard Business Review*, November 2004.

8. Eric Wahlgren, "Fast, Fashionable and Profitable: The Performance of European 'Cheap Chic' Chains H&M and Inditex Is Making Rival U.S. Apparel Chains Look Like Wet Rags," *BusinessWeek Online*, March 10, 2005.

9. Nikos Kardassis, interview by author, June 21, 2005.

10. Mike Hughlett, "Motorola Curbs Chase for Patents," *Chicago Tribune*, August 21, 2005.

11. Teruaki Aoki, interview by author, June 7, 2005.

12. Sony Corporation, "Sony Takes Home Three Emmy Awards for Technical Achievement," news release, October 17, 2001.

13. "R&D Scorecard Global Top 1,000 Companies: U.S. Firms Dominate; 86% of Total R&D Comes from Just 6 Countries out of 36," Finfacts Ireland, October 24, 2005, http://www.finfacts.com/irelandbusinessnews/publish/article_10003718.shtml.

14. Steve Luczo, interview by author, October 17, 2006.

15. "Inside Intel," *BusinessWeek*, January 9, 2006.

16. "Polaroid Introduces Its Digital Camera for the Midrange Market," *Wall Street Journal*, March 12, 1996.

17. James P. Andrew and Harold L. Sirkin, "Innovating for Cash," *Harvard Business Review*, September 2003.

18. Alex Taylor III, "The Birth of the Prius," *Fortune*, February 24, 2006; and David Welch and Lorraine Woellert, "The Eco-Cars," *BusinessWeek*, August 14, 2000.

Five: The Orchestrator

1. Bill Mitchell, interview by author, June 28, 2005.

2. Neil Fiske, multiple interviews by author team, 2004–2005.

3. Martin Ertl, interview by author, January 19, 2006.

4. Michael Mecham, "Betting on Suppliers," *Aviation Week*, October 27, 2003.

5. Michael Mecham, "Boeing Bets on Long Haul," *Aviation Week*, December 22, 2003.

6. Geoffrey Thomas, "Cruising at the Speed of Money," *Air Transport World*, April 1, 2003.

7. Mecham, "Betting on Suppliers."

8. David Bowermaster, "'Heavies' Help Carry 787," *Seattle Times*, May 2, 2005.

9. Daniel Michaels and J. Lynn Lunsford, "Planes, Trained by Automobiles," *Wall Street Journal Europe*, April 1, 2005.

10. Bowermaster, "'Heavies' Help Carry 787.'"

11. J. Lynn Lunsford, Daniel Michaels, Neil King, and Scott Miller, "New Friction Puts Airbus, Boeing on Course for Fresh Trade Battle," *Wall Street Journal*, June 1, 2004.

12. Roger Parloff, "Not Exactly Counterfeit," *Fortune*, May 1, 2006.

13. Mitchell, interview by author.

14. Todd Starr, interview by author, October 27, 2005.

15. Tom Arent, interview by author, February 6, 2003.

16. "Event Brief of Q4 2002 Whirlpool Corporation Earnings Conference Call," CCBN and FOCH e-Media, February 5, 2003.

17. Starr, interview by author.

18. "Event Brief of Q4 2002 Whirlpool Corporation Earnings Conference Call," CBN and FOCH e-Media, February 5, 2003.

19. Starr, interview by author.

20. "Sony Begins to Bear Fruit at Ericsson Venture," *Nikkei Report*, October 20, 2003.

21. E-mail from Gerald P. Cavanagh, Sony Corporation, June 28, 2006.

22. Ad Huijser, interview by author, March 15, 2005.

23. Judith Crown and Glenn Coleman, "The Fall of Schwinn," *Crain's Chicago Business*, October 4, 1993.

Six: The Licensor

1. Nicholas Varchaver, "A Hot Stock's Dirty Secret," *Fortune*, July 9, 2001.

2. Don Clark, "Rambus, Infineon Reach Settlement," *Wall Street Journal*, March 22, 2005.

3. Jamie Huckbody, "Pierre Cardin, He's Everywhere," http://www.theage.com.au/articles/2003/08/01/1059480531338.html.

4. Michael Maloney, e-mail to author, July 31, 2006.

5. Pamela Hawkins Williams, Dotcy Isom III, and Tiffini D. Smith-Peaches, "A Profile of Dolby Laboratories: An Effective Model for Leveraging Intellectual Property," http://www.law.northwestern.edu/journals/njtip/v2/n1/4/.

6. Siegfried Dais, interview by author, January 17, 2006.

7. "Reorganizing to Innovate: Procter & Gamble's Jeff Weedman," Yet2.com, http://www.yet2.com/app/insight/insight/20000917_weedman.

8. Jeff Weedman, interview by author, November 2, 2005.

9. Jim O'Connor, interview by author, May 12, 2005.

10. Ad Huijser, interview by author, March 15, 2005.

11. Andreas Gutsch, interview by author, January 31, 2006.

12. James Aley and Ann Harrington, "Heads We Win, Tails We Win," *Fortune*, March 3, 2003.

13. Quentin Hardy, "QUALCOMM Aims for Bigger Payoffs in Wireless," *Wall Street Journal*, January 21, 1999.

14. Dave Mock, "The Early Days of Cellular CDMA: Excerpts from a Discussion with Irwin Mark Jacobs, CEO of QUALCOMM," *The QUALCOMM Equation*, October 20, 2003, http://www.thequalcommequation.com/interviews2.shtml.

15. Ibid.

16. Steven Brull and William Echikson, "QUALCOMM: From Wireless to Phone-less," *BusinessWeek*, December 6, 1999.

17. Aley and Harrington, "Heads We Win, Tails We Win."

18. Brull and Echikson, "QUALCOMM."

19. Hardy, "QUALCOMM Aims for Bigger Payoffs in Wireless."

20. Brull and Echikson, "QUALCOMM."

21. "The Arms Race," *Economist*, October 22, 2005.

22. Ibid.

23. Ibid.

Seven: Aligning

1. "Citigroup Reshuffles Top Management Ranks," *Reuters News*, July 25, 2000.

2. Claus Friis, interview by author team, September 22, 2003.

3. Bartolomeu Sapiensa, interview by author team, October 27, 2005.

4. Elcio Pereira, interview by author team, November 11, 2005.

5. Friis, interview by author team.

6. Sapiensa; Pereira, interview by author.

7. Ibid.

8. Ibid.

9. Ibid.

10. Ibid.

11. Friis, interview by author team.

12. Pereira, interview by author team.

13. Ibid.

14. Ibid.

15. Jim O'Connor, interview by author, May 12, 2005.

16. Doh-Seok Choi, interview by author, July 10, 2006.

17. Jean-Louis Ricaud, interview by author, January 16, 2006.

18. Siegfried Dais, interview by author, January 17, 2006.

19. John Markoff, "In Race to Develop Blue Lasers, Japanese Star Surges Ahead," *New York Times*, January 19, 1999.

20. Yoshiko Hara, "Bright Blue LED Could Enable Color Displays," *Electronic Engineering Times*, March 21, 1994; and "Interview: Blue LED Inventor Looks to Revolutionize Lighting," *Nikkei Report*, July 31, 2002.

21. Teruaki Aoki, interview by author, June 7, 2005.

22. Karim R. Lakhani, Lars Bo Jeppesen, Peter A. Lohse, and Jill A. Panetta,

"Solving Scientific Problems by Broadcasting Them to Diverse Solvers," working paper 10, Harvard Business School, Boston.

23. Erwin Schrodinger, *What Is Life? The Physical Aspect of the Living Cell* (Cambridge, UK: Cambridge University Press, 1951).

24. Dais, interview by author.

25. Choi, interview by author.

26. Young-Ha Lee, interview by author, January 11, 2006.

27. "The Awakening of Qualia," http://en.wikipedia.org/wiki/User: Vuara/The_Awakening_of_Qualia.

28. Makoto Kogure, interview by author, June 8, 2005.

29. Karl Weinberger, interview by author, January 17, 2006.

30. Jeff Weedman, interview by author, November 2, 2005.

31. For more on metrics, see Boston Consulting Group, *Measuring Innovation 2006 Senior Management Survey*, 2006.

32. Claus Weyrich, interview by author, March 14, 2006.

33. Doh-Seok Choi, interview by author, January 10, 2006.

34. Ad Huijser, interview by author, March 15, 2005.

35. Martin Ertl, interview by author, January 19, 2006.

Eight: Leading

1. Larry Culp, interview by author, March 7, 2006.

2. Ibid.

3. Steve Luczo, interview by author, October 17, 2005.

4. Holger Schmidt, interview by author, January 17, 2006.

5. Doh-Seok Choi, interview by author, January 10, 2006.

6. Claus Weyrich, interview by author, March 14, 2006.

7. Tom Buckleitner, interview by author, September 20, 2005.

8. Didier Roux, interview by author, January 17, 2006.

9. Bill Mitchell, interview by author, June 28, 2005.

10. Weyrich, interview by author, March 14, 2006.

11. Kim Winser, interview by author, February 3, 2006.

12. Mike Snyder, interview by author, July 8, 2005.

13. Andreas Gutsch, interview by author, January 31, 2006.

14. Teruaki Aoki, interview by author, June 7, 2005.

15. Makoto Kogure, interview by author, June 8, 2005.

16. Yves Dubreil, interview by author, January 16, 2006.

17. Snyder, interview by author.

18. Young-Soo Song, interview by author, January 13, 2006.

19. Pattye Moore, interview by author, June 13, 2005.

20. Peter Ottenbruch, interview by author, January 21, 2006.

21. Dubreil, interview by author.

James P. Andrew is a Senior Vice President and Director of The Boston Consulting Group and heads the firm's global innovation practice. Jim works with leading companies around the world and across industries to develop innovation strategies, align organizations to enhance their culture and innovativeness, create breakthrough new businesses, redesign new product development processes, improve R&D management, optimize product portfolios, and design innovation metrics systems.

Jim and his ideas have been featured in dozens of leading international publications around the world. He leads BCG's annual global Senior Executive Survey on Innovation (conducted in conjunction with *BusinessWeek*), and is the lead author of the *Harvard Business Review* article "Innovating for Cash".

Jim joined BCG in 1986, founded and led the firm's offices in Mumbai (Bombay) and Singapore, and is now based in Chicago. He holds an MBA from Harvard University Graduate School of Business, with distinction, and a BS from the University of Illinois, with highest honors.

Harold L. Sirkin is a Senior Vice President and Director in BCG's Chicago office. He leads the firm's Global Operations Practice. Under his leadership, the firm has emerged as the foremost driver of client results in two areas critical to profitable growth—innovation and globalization. Hal previously led BCG's highly successful E-commerce and IT Practices.

Hal works with leading companies worldwide to improve their innovation returns, operating efficiency, global competitiveness, and strategic use of IT. His expertise spans a broad range of industries, topics, and geographies. A thought leader both within and outside the firm, he is frequently quoted in the press worldwide and writes a quarterly column for *BusinessWeek* on-line. He has authored a wide range of articles for business publications, including several for the *Harvard Business Review*.

Hal has been with BCG for twenty-five years. He holds an MBA from the University of Chicago and a BS summa cum laude from the Wharton School. He is a Certified Public Accountant.